AI and Humanity

AI and Humanity

Illah Reza Nourbakhsh and Jennifer Keating

The MIT Press
Cambridge, Massachusetts
London, England

This book was set in Stone Serif and Stone Sans by Westchester Publishing Services. Printed and bound in the United States of America.

Library of Congress Cataloging-in-Publication Data

Names: Nourbakhsh, Illah Reza, 1970- author. | Keating, Jennifer, author.
Title: AI and humanity / Illah Reza Nourbakhsh, Jennifer Keating.
Description: Cambridge, Massachusetts : The MIT Press, 2020. | "This work began as an
    experimental undergraduate course, AI & Humanity, offered in the School of Computer
    Science and the Dietrich College of Humanities and Social Sciences at Carnegie Mellon
    University"--Preface. | Includes bibliographical references and index.
Identifiers: LCCN 2019029886 | ISBN 9780262043847 (hardcover) |
    ISBN 9780262358156 (ebook)
Subjects: LCSH: Artificial intelligence--Philosophy. | Artificial intelligence--Social aspects. |
    Technology--Social aspects.
Classification: LCC Q334.7 .N68 2020 | DDC 303.48/34--dc23
LC record available at https://lccn.loc.gov/2019029886

10 9 8 7 6 5 4 3 2 1

For Mom. You have taught me to believe in the connections that I see.
—JTK

To Nikou, Mitra, and the CREATE Lab at Carnegie Mellon's Robotics Institute.
You inspire me every instant.
—IRN

Contents

Preface

This is not a TED Talk. In this book, you will find that our questions increasingly lead to more questions rather than to definitive answers. As we work through this thought experiment on the myriad relationships between developing artificial intelligence (AI) and humanity, we require your patience (mixed with generous doses of imagination!). We will introduce and explore the historical development of various AI systems and tools, moving from the seemingly innocuous Roomba to the considerably more sinister Harpy semiautonomous weapon system. We will consider the evolution of various AI tools and consider how these tools affect our individual relationships to systems and the ways in which these tools might impact relationships between humans (as individuals and collectives). We are at the beginning of what looks like a revolutionary period in relationships between humans and machines. How then might we anticipate the questions that will come to define this epoch? Will AI systems undermine our abilities to make decisions? Will these tools assist us in achieving scientific discoveries, health interventions, or relief of inequality beyond our wildest dreams?

We are only at the beginning of understanding what the features of this new era will be. But how might we better prepare citizens the world over with knowledge of the technological tools that continue to infiltrate and influence our lives? And what lessons might we learn from history to better prepare ourselves for the seemingly inevitable ruptures and changes already unfolding in this technological revolution? How can we anticipate certain cultural shifts and changes that might already be underway? And what tools will we need to regulate these changes to harness the good, rather than subject ourselves to the potential ill effects, of a rapidly changing, technologically integrated world?

Throughout this book, you will engage with two authors: a humanist, with educational training in literature and cultural studies, and a roboticist, educated as a computer scientist. Our disciplinary perspectives frame the course of our inquiry and serve as footholds for disrupting many mainstream discussions presently unfolding in regard to developing AI: its potential ethical implications, social impact, and cultural influence. This work began as an experimental undergraduate course, AI & Humanity, offered in the School of Computer Science and the Dietrich College of Humanities and Social Sciences at Carnegie Mellon University. It is one of a fleet of experimental courses, crossing conventional academic boundaries to attend to grand societal challenges. Through the lenses of literature, film, visual art, and memoir, students explore how people negotiate power structures and the ways in which individuals are influenced by emerging technologies in a variety of cultural and historical contexts. We consider how human relationships shape and influence the tools that we build and how these tools in turn shape how we navigate these relationships in society. For example, we consider the cotton gin. How did this technology influence demands for swaths of land and an expanded labor force, as well as contribute to an American corner on the international cotton market in the late seventeenth and early eighteenth centuries? How might this example compare to the advent of "thinking" machines in the example offered by the IBM Watson platform or issues pertaining to features of orientation and identification in developing semiautonomous vehicles?

We suggest the downstream impact of technological innovation on humanity only grows stronger as the tools that we use become increasingly sophisticated and integrated with modern life. As we engage with students who will become the leaders of future generations—technologists and humanists who will build, regulate, and critically consider the role of these tools—we are invested in building our readers' and our students' skill in framing these pertinent questions, narrowing the scope of the problems and identifying tools that will prepare us for navigating the coming chapter in human history. We are invested in building a common language to consider our shared history and in using this lexicon as a tool to analyze how rapidly advancing technological systems influence our lives.

Historically, schooling systems and universities have been poorly designed for the cross-disciplinary demands that require artful navigation of social and cultural questions. In this book, we consider how technological tools are devised and evolved to attend to specific questions and problems in society. We also interrogate how carefully we anticipate the intended uses or unintended uses of these systems as designers, engineers, programmers, consumers, and regulators. What happens when a faucet sensor is not attuned to variations in skin pigment and cannot "see" hands that fall beyond the spectrum of its programmed skin pigment recognition algorithm? How are drones coded to "identify" nonstate actors in a war zone? What features distinguish a combatant from a civilian in such circumstances? How does the language we use to describe AI systems that assist humans or even outpace our abilities use literary tropes or metaphor to imply a hyperbolic level of capability?

Our vision in this book is to equip readers with a robust vocabulary and fluency of technological advancement to ensure comfort and dexterity in attending to the emerging questions of our time. We want to help build a shared intellectual space that can be combined with structured opportunities to consider the world as it changes and to envision a world in which AI systems are harnessed for the betterment of society, rather than perpetuating its ills. Today we inhabit a boundary space where we test the agility of our social and political conventions. We face a wave of revolutions last seen during the rise of industrialization in the nineteenth century, the machine age at the turn of the twentieth century, and the advent of the digital economy in the late twentieth century. Like in recent histories, our political and social systems are not equipped to attend to the impact of these systems on human populations—and certainly not to the influences on our most vulnerable individuals and communities worldwide. Our educational institutions are in a race to catch up to the reality that is already knocking at our doors but has been predicted for decades.

In 1965, British mathematician I. J. Good wrote, "An ultraintelligent machine could design even better machines; there would then unquestionably be an 'intelligence explosion,' and the intelligence of man would be left far behind."[1] As we enter an age in which companies like Uber, Argo, and Aurora are testing driverless cars in Pittsburgh and innovative interfaces like IBM's Watson can play *Jeopardy!* and learn techniques for medical diagnoses, we are negotiating both how we navigate this intelligence explosion and how these explorations influence our understanding and perhaps preservation of what it means to be human. The future of human-machine relations will likely define our historical epoch, and yet many young technologists and humanists underestimate the downstream impact of technological innovations on human society.[2] Our leading experts are also scrambling to prepare for a future that has begun before it was expected. We grasp at metaphors for language to describe these systems and what we expect they can or will do in an immediate or far-flung future. Popular reports betray a lag between the dreams associated with these systems' ultimate functionality and their present levels of efficacy. Uber may describe a near future in which no one needs to drive and accidents reduce to naught, but we know that as of summer 2019, the systems are often "timid"—or worse, incapable of making the decisions needed to safely navigate populated roads in cities like Pittsburgh, San Francisco, or Tempe.[3] To positively influence the future arc of AI and humanity, we must study human-machine relations in the context of our human past and our technology's possible future. In this book, we offer that very practice to be repeated in dynamic ways by our readers: curious citizens, university students, or future technologists and inventors.

Acknowledgments

This book would not exist but for the encouragement offered by Marie Lufkin Lee at MIT Press. Marie sat with us, listened to the story of our experimental class, and proposed that we convert our lectures into a book. Her sound advice and vision in regard to this teaching project provided the fuel that we needed to dedicate two years to this collection of essays. The students in our inaugural AI and Humanity seminars were also essential to this effort; their conversations, questions, and explorations were catalysts that helped this work to mature from ideas into the more carefully framed inquiries that you will find in this book.

Our readers provided invaluable guidance to take this book from a ragged, early concept piece to finished form: Professor James Knapp and Professor Peggy Knapp, Professor Michael Genesereth, Maureen Bard, Professor Colin MacCabe, and Steven Ketchpel offered close readings of early drafts. Amy Burke performed comprehensive image rights research; Mitra Nourbakhsh undertook early transcription work. Thanks to Sophie Smith for exhaustively transcribing nearly all of the AI and Humanity Oral Archive and to Sophie and Joon Jang for creating the course website at aiandhumanity.org. Brian Staszel elegantly compensated for our inadequacies as videographers during video postproduction. Our colleagues at the Dietrich College of Humanities and Social Sciences, including Dean Richard Scheines, Professor David Danks, and Associate Dean Brian Junker, offered conversations that informed our respective thinking. At the Robotics Institute and in the School of Computer Science, Professor Reid Simmons, Professor Henny Admoni, and the CREATE Lab offered help in sharpening the ideas captured in this book. Our respective colleges offered the time, support, and space to pilot our teaching efforts with students in our respective schools. Our thanks go to our colleagues in the dean's offices of both schools—namely, Associate Dean Guy Blelloch, Assistant Dean Tom Cortina, and Dean Richard Scheines—for their support in this curricular experimentation.

We are thankful for Stephanie Cohen's assistance in our production process at MIT. We are also thankful for Professor John Carson's collegiality and contribution to cover art. We also thank Rita Duffy for her inspirational work. Finally, we thank our families for putting up with the time that we continue to invest in our respective academic enterprises.

1 Introduction

Figure 1.1
Swiss automaton writing letters longhand, circa 1770.

The immediate question that was in my mind is, is this different from when owners of stables started to lose their jobs as horses were replaced with cars.
—Andrew Moore, Google, former dean of the School of Computer Science, Carnegie Mellon University[1]

But the question is, is it more or less than the death that would happen if it was just a person who was driving the car, right? … Is death more acceptable when a human being does it than when a machine does it?
—Louis Chude-Sokei, professor of English, chair of African American Studies, Boston University[2]

People talk a lot about oversight, humans having oversight over AI, but if you really go back and think about it, who's making more mistakes, humans or AI? Well, I would argue it's humans are making a lot more mistakes than AI, so maybe there should be AI oversight over humans and maybe it should be a little bit of both ways.
—Tuomas Sandholm, professor of computer science, Carnegie Mellon University[3]

I'm a great fan of driverless car technology, it's just that we're not doing it right.
—Alan Winfield, professor of robot ethics, University of the West of England[4]

This textbook offers an analytical framing and a developing language for identifying, describing, and attending to the influences of contemporary technological advancement on societal structures and customs. Our uncertain future will be shaped and influenced by the advancements of the coming two decades, a glimpse of which we can already see today. As we build skills to shape and ensure responsiveness to this developing future, this book offers techniques for building a common language to describe and analyze the future that is now unfolding. Together we will attend to questions pertaining to individual and group identity, the influence of technological innovation on society over time, and explorations of the possible areas in which the next generation of thinkers is likely to need clarity and confidence to safeguard our societies. Our readers, students, and communities will develop policies, laws, and expectations for participating in this near future, protecting the features of human society that we most value.

The diversity and dynamism of our readers, combined with the boundary content of this book, make semantics and language a foundational aspect of this shared work. Words such as *autonomy*, *agency*, *technology*, and *identity* are threatened with reformulation as artificial intelligence shifts our understanding of what it means to be human and what it means for a machine to behave in human-like ways. For humanists and technologists, these words may lack the comprehensive etymological context that can help to situate discussions regarding new technology. Just as importantly, precise language is a crucial tool to help identify parallels between new technological systems that shift or change user and machine relationships and old negotiations of power. This intertwining of a historical relationship with evolving technological tools, their influence on human relationships, and analyses of how these past examples equip us to navigate the unfolding future will be our joint exploration in *AI and Humanity*.

The structural backbone of this book is steeped in the etymology of critical terms, drawing attention to the circulation and evolution of the English language over time. Using *Keywords* by Raymond Williams and *Keywords for Today* by Colin MacCabe and Holly Yanacek as foundational texts, we explore the cultural and etymological roots of boundary words relevant to AI and society. These terms drive forward an ever-expanding common language that is used to evaluate historical and future-facing explorations of technology and humanity, ranging from plays like Karl Čapek's *RUR (Rossum's Universal Robots)* to memoirs like Frederick Douglass's *Narrative of the Life of Frederick Douglass, an American Slave*.

Although language is the foundation of our shared work, we will delve into the detailed history of technological advances that have given rise to AI and use literary explication and analysis as the lens for exploring the societal impact of these systems. A single disciplinary perspective simply cannot capture or frame the myriad challenges that face us now and in the coming decades. Several generations ago, C. P. Snow's "The Two Cultures" (delivered as a Rede Lecture at Cambridge University in 1959) offered an indication of how disciplinary specialization might serve as an impediment to attending to society's greatest challenges. As he considered the role of practitioners in the arts and sciences, broadly conceived, Snow claimed that warring factions on either side of a divide were so susceptible to partisanship that they would ignore the pressing issues of the age. He states: "If the scientists have the future in their bones, then the traditional culture responds by wishing the future did not exist. It is the traditional culture, to an extent remarkably little diminished by the emergence of the scientific one, which manages the western world. The polarization is sheer loss to us all. To us as people, and to our society. It is at the same time practical and intellectual and creative loss, and I repeat that it is false to imagine that those three considerations are clearly separable."[5]

Attending to the intertwining of disciplinary divisions that scale to societal implications and shortcomings, Snow's words betray echoes of the Cold War era. His casing of the scientist as diametrically opposed to the humanist or "traditional culture" predicts our current state of affairs, in which technologists are "king" and a cult of reason justifies privileging the scientist, engineer, and technologist over all others. Snow's predictions arguably dominate our economic and political reality today.

Presently, the prospect of AI-based, global deployments threatens to shift entire work categories from manual to autonomous, from milling machines and sports journalism to the entire category of the driven machine: forklifts, taxis, buses, trucks, and private automobiles. If ever AI were absent from societal impact because of its station on the blackboard, like theoretical mathematics before the nuclear age, that epoch is now history. Billions of dollars are now in play because venture capitalists, the most financially rational of all agents, believe that they will reap significant financial gains from real deployments that are only five years away by their yardstick—gains that will signal massive shifts in labor-to-capital wealth and from human skilling to machine autonomy and agency. Politicians the world over are bending to this emerging power structure in which capital reigns supreme, driven by tech industries. But as we educate the next generation of technologists that will drive this power structure, how might we engage technologists and humanists alike to foster understanding and perhaps fashion a shared investment in protecting some of the most delicate characteristics of human society?

In his Rede Lecture, Snow offers a timely intervention to the divide that he aptly predicts in the "traditional culture," which eventually translates to representatives from the academe in the humanities and arts traditions and their technically oriented brothers and sisters. "There is one way out of this," he writes. "It is, of course, by rethinking our education."[6] In this spirit, we strive to fashion bridges between humanities and technically oriented citizens and students in classrooms and discussion circles. We devise a set of analytical skills and interdisciplinary practices that encourage you to actively engage and shape ethical and conscientious relationships between human society and the tools that we fashion for navigating our world.

AI and Humanity attends to existential questions about what it means to be human (read: not machine) in the context of a rapidly advancing technological age. We consider human narratives throughout history that examine how governments and individual citizens defined humanity in the context of slavery and colonialism as a framework for exploring and projecting what it means to be human in the age of rapidly advancing "machine intelligence." We trace the technological advancements of the most recent five decades and identify historical precedents and speculative narratives that help us to consider issues like labor, economic disparity, negotiations of power, human dignity, and ethical responsibility within the context of human relations and the advancing wave front of intelligent robots. Within this textbook, you will study historical texts, modern articles, documentaries, and science fiction. You will synthesize models and conceptual narratives that interrogate our shared future with artificial intelligence in light of our past struggles with power and agency. You will be introduced to exercises like concept mapping, journal entries, and creative projects that our students have undertaken with success. Their example will serve as instigators for further discussion in your own local circles.

Our learning goals in the undergraduate course span contextual analysis of human history and future-facing explorations of the possible ramifications of present-day AI innovation. Their relevance to this book will become clear in the pairing of content with suggestions for further discussion or investigation. Our goals include the following:

1. Identify, describe, and respond to historical examples of negotiations of power between human individuals and communities.

2. Develop language to describe and evaluate the historical and contemporary evolution of machines and of human relationships to these systems.

3. Survey a variety of narrative forms that explore human relationships to emerging technologies over time, including futurism.

4. Map foundational technology innovations that have resulted in and might lead to disruptive advancements in artificial intelligence.

5. Create individual and collaborative narratives pertaining to the evolving relationships between humans and machines.

As we build precise language, we wish to encourage enthusiasm for inquiry and a value perspective that reformulates each reader's relationship to AI-related technology. We strive to build sensitivities to the influences of AI on society over time, to acknowledge the immediacy and particularity of issues that will face communities today and in the near future, and to draw on the past for lessons learned or parallel concerns. To the readers who will determine our collective future, AI can no longer be a moniker for unguarded, unfamiliar technological innovation. Instead, AI should be intimately understood as a boundary development that is comprehensible and decipherable, a set of tools to empower individuals rather than myriad alternatives. We expect our readers to become advocates for responsible social integration of technology by every citizen in everyday life rather than to uphold values that lead humans to become subject to technological power. As readers and responsible citizens, you are the vanguard of such an active, value-oriented community in fast-changing times. As you build the next generation of tools as technologists and decide to analyze its range of influences on our present and future society, you will write our next-generation narrative.

Although keywords drive this inquiry, they comprise one of seven broad categories of content we employ: language, history, cultural analysis, art, futuring, tech notes, and news analyses. In documentary films like Herzog's *Lo and Behold*, we consider meditation on the ramifications of the internet on social relations in contemporary society. Simon Head's *Mindless* describes working conditions under surveillance and the threat of automation at Amazon and Walmart. We access history to elucidate negotiations of power, agency, and identity, with a close reading of Frederick Douglass's *Narrative of the Life of Frederick Douglass, an American Slave*. Philosophical texts expose readers to a framework for defining the self and identifying characteristics of narrative. Rita Duffy's paintings offer creative and critical commentary on the role of surveillance technology on society, prompting readers to consider their own possible futuring projects or thought experiments. Works of futuring in science fiction provide a fertile field for synthesizing keywords, potential ramifications, or perceived implications for specific technological advancements in discrete historic periods that range from the 1920s, with Čapek's *RUR*, to the 1980s with *Star Trek: The Next Generation*, the 1990s with *Minority Report*, and the early 2000s with *Black Mirror* episodes. We cite work and interleave interview transcripts from globally eminent scholars using the AI and Humanity

Archive as we attend to themes that range from autonomous weaponry, human identity in the age of AI, and ethics and AI to technology primers on state-of-the-art innovation. Each chapter's technical context provides primers on the specific history of
technology development and state-of-the-art innovations relevant to each keyword
group. News articles are introduced along specific keyword themes to demonstrate contemporary rhetorical slippage in the ways that scientists, politicians, and other stakeholders describe and articulate the significance of our rapidly developing systems.

As a tool to catalyze the development of dynamic discourse, materials such as journal entry prompts, discussion questions, and concept map challenges are included for
each chapter. Assignments in the course, and in each chapter of the book, are crafted
through the analytical lens of keywords sets. Two journaling assignments prompt
reflections on the relationship you as an individual craft with developing technology and the community to which you belong. We ask you as readers to identify the
ways in which the book's content and exercises influence your own thought processes
regarding the role of AI in society and the active role that you might play in the coming changes ahead of us. This exercise facilitates your consideration of *ontology*, the
structure of concepts, as presented in a specific piece of work, but also in relation to the
clusters of keywords corresponding to chapter themes. This is an opportunity for you
to undertake the interpretation of language to complement the analysis that we offer
in each chapter.

This field of inquiry is dynamic, and it is important for this text to be a window into
others' analyses. It is also a vantage point to consider current events. Our companion
website, *aiandhumanity.org*, provides up-to-date resources related to all assignments in
this book. It also provides a portal into the AI Oral Archive Project, in which we have
collected remarks from AI scientists, sociologists, literary theorists, and others, to provide the public with broad exposure to expert thinking in regard to artificial intelligence at present.

The coming decades will prove a significant, transitional period in which it is difficult to predict or even attend to the lived experience of societal rupture and building
anew. As we witness breathtaking advances in AI systems, we rush to grapple with
rapidly evolving state-of-the-art systems and our imagined and lived understanding
of their influence on human and machine relationships. This book serves as a tool to
equip you in this evolving process—a set of exercises to be visited and revisited with
new materials, new lines of inquiry that are still linked to a relatively stable grasp of the
past. As we look toward the future, a moment of pause and reflection on where we have
come from will serve us well so that we do not make the same mistakes as our forebears.

Cautious optimism can help us build a shared future that protects our most vulnerable and equips all citizens with the tools to thrive.

Discussion Questions

1. How is AI portrayed today in public discourse? How is it represented as a threat, and how is it offered as a savior for society?
2. What historical technologies have influenced our human cultures most profoundly?
3. What anxieties do you have regarding the role of technology such as AI and computational systems in our shared future?
4. How does computing technology play a role in your expression of your personal identity?

2 Technology and Society

Figure 2.1

Personally I think the idea that fake news on Facebook … influenced the election in any way is a pretty crazy idea. Voters make decisions based on their lived experience.

—Mark Zuckerberg, 2016[1]

We did all the research, all the data, all the analytics, all the targeting, we ran all the digital campaign, the television campaign and our data informed all the strategy. We just put information into the bloodstream of the Internet and then watch it grow, give it a little push every now and again over time to watch it take shape, and so this stuff infiltrates the online community, but with no branding, so it's unattributable, untrackable.

—Alexander Nix, Cambridge Analytica, 2018[2]

So this was a major breach of trust and I'm really sorry that this happened. You know we have a basic responsibility to protect people's data and if we can't do that then we don't deserve to have the opportunity to serve people.

—Mark Zuckerberg, 2018[3]

Sources

Lo and Behold, Reveries of the Connected World, a documentary directed
by Werner Herzog
Keywords: Society
Keywords for Today: Technology

Guiding Issues

This chapter begins our exploration of artificial intelligence and humanity by focus-
ing on two terms that calibrate and orient the book's trajectory: *technology* and *society*.
To have meaningful discourse, we must study both the roots of these two words and
the ways in which they interrelate. The influence of technology on society is of course
studied in depth: agricultural tools, currency, the systems for dispersing the written
word—all of these represent technologies that have utterly changed the dynamics of
societal evolution and the forging of communities worldwide. Here we focus, particularly, on modern technology. Both AI and robotics are new technologies that are the
twin children of digital computation. We study these newest forms of technology, AI
and robotics, because they pose wholly new challenges in our understanding of how
technology might influence society—and these new challenges constitute the heart of
our exploration. The history of another sibling to AI and robotics, the internet, serves
as a guide rail in the documentary film *Lo and Behold, Reveries of the Connected World* by
Werner Herzog.

    Before diving into an analysis of technology and society, consider what makes these
most modern technologies unique. Consider, first, artificial intelligence. In recent years,
AI progress has caught considerable momentum, pushing the discipline from a theoretical field of computer science and engineering study since 1960 into a broader,
developed, real-world disruption miles away from its beginnings in laboratory demonstrations. Discourse pertaining to AI no longer remains sequestered in universities, tech labs, or academic journals. The developing systems infiltrate every aspect of
contemporary lives, from curated news reports on iPhones to enlisting Alexa's help
in ordering groceries. The discourse pertaining to these systems and tools also dominates current news headlines and television coverage. Theoretical computer vision
techniques invented in the 1980s now operate in real time, enabling face detection,
emotion extraction, and countless other computational faculties for turning the analog world into digital artifacts that are ripe for autonomous, quantitative analysis and

response. Advances in AI-based sensing and machine learning have opened the possibility of dynamic physical systems under autonomous control: drones can already weave between trees in a forest and fly through open windows; running machines can outpace humans; sex robots have moved from fiction to start-up company warehouses. Tens of millions of Facebook accounts can be mined by AI algorithms to create digital marketing products that directly influence democratic election dynamics. But how well equipped are we to attend to these dramatic shifts in our society?

Herzog's film offers provocative insights as we consider the influence of technology on society. In addition, it suggests a need for understanding the nuanced semantics of two heavily laden words: *technology* and *society*. We begin this chapter with a study of the surprising roots of these words. We then consider the application of our semantics in the context of *Lo and Behold* and its treatment of the internet technology revolution. As we consider the full ramifications of modern AI innovation, we can also consider questions pertaining to its influence and ownership. How will modern technology influence our collective and individual identities in society? How will societies incorporate or adopt technology into specific evolving cultural practices? Will AI in fact redefine how we understand society?

Language: Technology and Society

In current discourse, *technology* has come to signify much more than tools that are devised based on cutting-edge scientific findings or advanced engineering principles. Instead, *technology* is often presented as a driving force, interwoven into political, cultural, and social elements of human society, likened to "progress" or "capital." It carries teleological undertones that often gesture toward utopian or dystopian futures, exhibiting features of human ambition and anxieties associated with existential threats. But historically, the term *technology* had little to do with science or engineering—until the post-Enlightenment era and mid-nineteenth-century industrial revolutions in the West. According to MacCabe and Yanacek in *Keywords for Today*:

> Technology was used from the seventeenth century to describe a systematic study of the arts or the terminology of a particular art. It is from fw teknologia, GK, and technolgia, Mod. L—a systematic treatment. The root is tekne, Gk—an art or craft. In the eighteenth century a characteristic definition of technology was "a description of arts, especially the Mechanical." It was mainly in the mid-nineteenth century that technology became fully specialized to the "practical arts"; this is also the period of technologist. The newly specialized sense of Science and scientist opened the way to a familiar modern distinction between knowledge (science) and its practical construction—and technological—often used in the same sense, but with the

residual sense (in logy) of systematic treatment. In fact, there is still room for a distinction between two words, with technique as a particular construction or method, and technology as a system of such means and methods; technological would then indicate the crucial systems in all production as distinct from specific "applications."[4]

A philological treatment of *technology* offers a historical record of the evolution of this term in English, including the linguistic influences of its Greek and Latin antecedents. This framing offers insights into the contemporary circulation and cache of *technology*, as well as its current mismatches in debates that often privilege science and technology over the humanities and arts.

With its origins as a "study of the arts or the terminology of a particular art," the circulation of *technology*, as a word, illustrates an ironic tension in our contemporary culture and future aspirations. In the West, advancing *technology* is often placed in rhetorical competition or positioned as a counterpart to the arts or humanities. Advancing technology is frequently associated with utopian visions for fully automated systems and universal incomes for citizens who might harness such tools and systems in all facets of individual lives and communities. The history of the term *technology*, however, suggests a need to access its history in the arts. In this history, we might find the very language needed to bridge gaps in communication between those who build and develop technological tools and those who are concerned with, and explore, their impact on society.

In *Keywords for Today*, the contemporary circulation of *technology* is captured:

> Although technology denoted "[a] discourse or treatise on an art or arts" from the early seventeenth century to the mid-nineteenth century, this sense is now obsolete, and many perceive art and technology as at odds with each other. Beginning in the late twentieth century, national science programs and education councils were established in the U.S. to promote excellence in the disciplines of science, technology, engineering and math (known as STEM since 2001). The perceived conflict between the liberal arts and a STEM education has been a contested topic of debate in the early twenty-first century. ... When introducing the iPad 2 in March 2011, Steve Jobs described Apple's strategy for success as one that integrates the arts with technology: "... technology alone is not enough—it's technology married with liberal arts, married with the humanities, that yields the results that make our heart sing."[5]

Tensions between privileging of the arts, humanities, and/or science throughout the West are captured in the etymology of *technology* and acknowledged in Jobs's jovial suggestion (whether deliberate or purely by accident). But the idea that intertwining of "liberal arts, married with the humanities" and technology can together "make our heart sing" is tempered again if we consider the context. Jobs's essential ambition, his utopian vision, was the ubiquitous intertwining of technology with every aspect of our lives to yield massive profits for Apple.

Although Jobs's chief concern is communicating the aesthetically pleasing and potentially ubiquitous tool encapsulated in the iPhone a decade ago, his notion of intertwining "liberal arts, married with the humanities" belies the significant age of technology infiltrating and deeply influencing human-to-human relations in our contemporary society. This suggests a significant shift in human-to-machine relationships too, which in turn affect human-to-human relationships. As we consider the specific tools that make up a current understanding of *technology*, we are compelled to consider also how these tools specifically impact past and present conceptions of human individuality and configurations of human communities. If technology has indeed infiltrated all aspects of cultural, political and social life in the West, what are the features of the society that is subject to such influence? What discernible characteristics in the terminology of today are indicated in the etymology of this term? And what systems concern us most as we consider how technology infiltrates defining features of society?

Society

In *Keywords*, Williams distinguishes two main "senses" for the term *society*: "the body of institutions and relationships within which a relatively large group of people live; and as our most abstract term for the condition in which such institutions and relationships are formed."[6] These distinct senses of *society* suggest a strategy for considering the manner in which technology can influence, implicate, or perhaps even alter relationships between individuals in the contemporary moment and the language that we use to describe such circumstances. But these working senses also suggest that we need evolving language to describe the scaled implications of advancing technology in the early twenty-first century. How is technology influencing the human "institutions and relationships" that are formed beyond individual "fellowship" that is suggested in the fourteenth century and sixteenth century manifestations of the term *society*? To what degree do we need to work on building a common language even within a single society or nation-state to better represent and perhaps understand the unique influence that technology might have on our sense of individuality and collectivity? By the nineteenth century, Williams states, "society can be seen clearly enough as an object to allow such formations as social reformer (although social was also used, and is still used, to describe personal company; cf. social life and social evening). At the same time, in seeing society as an object (the objective sum of our relationships) it was possible, in new ways, to define the relationship of man and society or the individual and society as a problem."[7]

Hetero-normative language aside, Williams's description of the manner in which society seen as "object (the objective sum or our relationships)" is a means for defining the relationship of "man" (read person) and society "or the individual and society as a problem" offers a useful baseline for considering the relationships between *technology* in its contemporary sense, as both a set of tools and a political, cultural, and social force, and *society*, as an object and as an abstract set of human relationships understood as a dynamic collective. When we think of these terms as separate, we can consider the manner in which the language serves as a symbol for the signified object(s). When we consider the interplay of technology as an actor on a narrow or broad definition of society, we then begin to touch on the intellectual sensitivity and power of this topic, which can be explored in any variety of ways in the classroom or in everyday discourse. As the sophisticated AI systems that develop now and in the near future weave into our lives, the significance of a shared language to describe and regulate how these parts of our world interface will become less of an intellectual exercise and more of a necessity.

The features of an exploration of the interplay between technology and society are far-reaching and offer considerable range. The outline presented in the subsequent chapters opens an investigation that touches on specific features of contemporary ranges of technological tools and the social and political issues that emerge as they are considered. Our terms, introduced in sets throughout the subsequent chapters, bring together specific facets of technology with particular forces pertaining to the identity and evolution of society, predominantly in the West. Our key concepts include the following:

- Labor, inequality, and dignity
- Narrative, exploitation, and labor
- Agency and autonomy
- Agency, labor, and narrative
- Surveillance and agency
- Literacy and data
- Data and individuality
- Autonomy, surveillance, and weaponry

As we explore the features of specific tools within the category of contemporary technology, we begin to delve into the particular influences and concerns for society in the present day. We can also determine the degree to which human history offers clues and guiding principles for how best to anticipate and guard against potential ills as we analyze the work of artists, writers, and technologists who explore these ideas.

Machine Rapture: The Evolution of Internet Technology

The evolution of the internet is an excellent study in the analysis of technology as a form of apparent progress that touches both senses of *society*. It is a system that both redefines networks of relationships among individuals and creates the matrix in which future societal evolution occurs. The technical foundations of machine-to-machine communication were laid as early as the 1960s, thanks to US Department of Defense and NASA innovation. Although those early efforts were slow to influence society, Herzog captures them brilliantly with the story of the first transcontinental (hel)lo message. The first attempted internet message, "Hello," only succeeded partially after the first three letters failed to transmit properly, resulting in "lo," as adopted by Herzog in his documentary title. Computational technology's roots—its metaphorical etymology—is laid out in *Lo and Behold* as the desire to communicate. And this attends significantly to the reality of computational technology's evolution over the past decades. More than any other force, it is the charge to communicate in more sophisticated ways with our fellow humans and, thereafter, with our machine companions, that has pushed technology into uncharted waters.

How interaction itself—interaction between machines, machine and human—has evolved over time constitutes a powerful narrative for understanding a technology that now strongly influences society. A brief review of the evolution of these computing interfaces will uncover specific trends in machine-human relations.[8] Early computers, as shown in *Lo and Behold*, initially used textual interfaces to accept commands and display results. The language between computer and human user was syntactic and symbolic, and the pace of human-machine communication was termed *single-threaded*. That is, the user would type a single command in, then await a single response back from the computer. So command and response were perfectly interleaved, and each communication act from the computer was a response to human interrogation: lines of power were clearly delineated to privilege the human master.

As computer interfaces evolved, most famously thanks to new design patterns arising from Xerox PARC's laboratories and maturing at Apple Computer and Microsoft, the computer screen evolved from a repository for symbolic, textual, master-slave dialogue to a two-dimensional graphical blackboard. With graphical representations of folders, trash cans, and other physical concepts, the computer screen shifted from syntax to metaphorical representation. The advent of windows arranged across a computer screen added a wholly new interaction paradigm: that of multiple instantiations of interaction, all occurring in parallel. In this graphically centered world, human users depend upon their sense of physical spatiality alongside symbolic understanding and furthermore can invoke multiple computer operations simultaneously across multiple *windows*, awaiting results from each window, in any presented order.[9]

The earlier interleaved, synchronized, master-slave model was replaced by new relationships: now computers no longer simply responded in lock step to human commands. Computers were busy with multiple activities and would notify humans of results episodically, as they happened to be available. The systems begin to echo the rise of society; machines relate to one another and to a myriad of human users. For humans, in turn, this shift in communication rhythm meant that the act of communicating with a computer was no longer a narrowly focused affair. You might launch a series of inquiries with the computer, grab some lunch, and later receive a result from the computer absent further inquiry. This was tantamount to a sea change in the relationship between human and machine because, unlike every well-designed industrial artifact that is only responsive, from a toaster to a coffeemaker, the computer became an object that could interrupt the ordinary flow of life activities with new information proactively: anytime, anywhere.

It is in the context of this shift in the human-machine relationship that the internet's birth further amplified a new human-machine mode. With the high spatial connectivity that the internet presented, it enabled both distant machines and distant humans to have direct impact on the knowledge and decisions of proximal communities. These advancements began a new logic of how society functions and communicates and the parties that compose that society.

Large knowledge and communication networks today support even further asynchrony, causing a shift in the dynamics of communication from a synchronized interaction between two individuals to unsynchronized, broadcast communications from many individual authors to communities of subscribers. The modern internet-connected machine, from televisions and mobile phones to refrigerators, is a small node in a highly connected system of knowledge and reaction that spans the world. Fundamentally, it supports chaotically asynchronous interactions across the network, between humans and between machines and humans. In doing so, it pervades the minute-by-minute activities of all who connect to the network.

Herzog documents the fully connected societal condition by revealing the sinister online harassment witnessed by two parents after the loss of their daughter, Nikki Catsauros.[10] The car wreck, a formerly private circumstance, is made at once public and personal by online trolls, who are able to search for, find, and communicate with objectionable imagery of the dead body. Although this starts at low-scale distribution after an officer makes the first bad choice of sharing an unfolding accident case with peers, the magnitude of scaled impact is evident with the viral distribution of these images.

Technological advancements in this case do not simply replace less efficient routines; the standard story of the typewriter replacing the fountain pen has astronomical effects in this case, carefully mined by Herzog. Instead, the internet invents new forms of knowledge seeking and communication—new forms that in very real ways can influence a family's process of grieving, coming to terms with tragedy and their unfolding relationships with communities near and far. Herzog's film mines these intended and unintended implications, creating an inquiry and critique that is "both eerily banal and breathtaking," connecting the dots between our technological advancements and the fabrics that they both weave and rend in their various uses.[11]

Looking even further forward, the Internet of Things revolution miniaturizes the final nodes of the existing global network to enable each and every made product to become a computing device. This may change the communicative relationship that consumers and users have with, literally, every thing. Your refrigerator, automobile, couch and home are all en route to become interactive devices that will maintain a social and interactive presence in your life. The final step in this communicative evolution is, of course, the wearable computing node. As the internet inexorably integrates more intimately with each human body, whether through implantation technology or clothing-based technology, we can expect the internet connectivity trend line to stop only when each person moves from organic, animate being to being both a living organism and a data-collection interaction point. As for power relationships, the Internet of Things completes a trend that started two decades ago. Humans are shifting from *task controllers*, initiating machines and then accepting the machine's output as its definitive master, to *network participants*. We are increasingly members of a heterogeneous network of people, physical devices, and machines, all of which can initiate contact, change goals, and reconfigure the network itself.

The role of the individual human in such a massive, global human-machine network remains an open question. It is one that we study in this book along axes such as surveillance, agency, and power negotiations. Without doubt, this conceptual reprogramming of individual identity is likely to influence our collective understanding of society: its makeup, its relative power, and its stability during a time of rapid technological revolution.

Societal Impact and Technology Today

In *Lo and Behold*, Mrs. Catsauros suggests that the internet "is the spirit of Evil, and I feel it is running through people."[12] She has viscerally felt the impact of this technology, and it has, in turn, become a palpable iteration of evil itself. Technology is ephemeral, however, still beyond her control. So the system can only be described in mythological or religious terms; it is outside her paradigm of human capability and function by virtue of its scaled reach and acute affect. The broader attitudes pertaining to how we make sense of technology, as demonstrated by Herzog, step back from this precipice that focuses on the machine and pivots toward another: human existentialism. Along multiple dimensions, Herzog's film shows how technology instigates possibilities for both provocation and salvation in relation to existential questions and crises in our social realities. The very format, composed of several vignettes addressing topics through the narratives ascribed to particular lives, attends to the fractures and attempts to coalesce tensions in our boundary moment in human history. By studying solar flares and the potential they constitute to destroy the internet communication infrastructure, Herzog shows how society's evolution has led us to paint ourselves into an existentially threatened corner, in which the entire system is highly fragile to elements beyond our control. The theme of existential risk is captured well by Ted Koppel in *Lights Out*, in which he demonstrates how a multiweek electricity outage can hastily beat society back to a dangerously primitive function.[13] But fragility is not the only message in Herzog's work.

The ultimate answer to the existential fragility of society becomes, ironically, an ever-increasing reliance on technology and the prospect of building somewhere new. Elon Musk espouses a Mars colony in *Lo and Behold*. As he quite seriously offers the establishment of an extraterrestrial society as our collective salvation from Earth's fragility, Musk betrays his South African roots. Colonial systems, empire building: Are these new motifs in human history? No. But haven't we learned from the countless downfalls of colonial enterprises the world over? Have we forgotten the lessons offered through Marlow in *Heart of Darkness*?

> They were conquerors, and for that you want only brute force—nothing to boast of, when you have it, since your strength is just an accident arising from the weakness of others. They grabbed what they could get for the sake of what was to be got. It was just robbery with violence, aggravated murder on a great scale, and men going at it blind—as is very proper for those who tackle a darkness. The conquest of the earth, which mostly means the taking it away from those who have a different complexion or slightly flatter noses than ourselves, is not a pretty thing when you look into it too much. What redeems it is the idea only. An idea at the back of it; not a sentimental pretense but an idea; and an unselfish belief in the idea— something you can set up, and bow down before, and offer a sacrifice to.[14]

For Musk, technology drives salvation. Unlike the colonial paradigm of the past, Mars offers a clean, uninhabited horizon. It is likened to Marlow's "blank space" on a map, but it is only available to those with the economic and political means to escape the binds of Earth for a specific form of salvation.

There is a second form of existential risk presented in *Lo and Behold* that is more nuanced in its ramifications on society than the colonial motif. It is the trope of *replacement* rather than global extinction. How does AI change society? And, specifically, how will the newest AI advances replace aspects of human function and perhaps features of our individual or collective identity? Here Herzog focuses on the reveries of technological optimism surrounding the self-driving car. In his interview with scientist Sebastian Thrun, we see a perfectly documented specimen of the early excitement surrounding the self-driving car. Thrun argues that humans are killers behind the wheel and that driving is a job that really should be left to the machines instead. He also offers a connected-world utopia, noting that the errors of self-driving cars transform into global fixes because of the hyperconnected future of these car networks: "Whenever a self-driving car makes a mistake, automatically all the other cars know about it, including future unborn cars."

Human replacement does not lead to a simple, one-to-one shift from human to machine in society. Rather, in this interpretation, the network of new driverless cars behaves nothing like the society of human drivers, but rather more like the fictional, networked Borg race.[15] Of course, you can temper this optimism with the recent history of driverless car technology: 2018 has witnessed multiple deaths that have shocked the public at large.[16] While technology companies, espousing Thrun's enthusiasm, ride at full tilt into the future, legislators desperately try to keep up. Unlike pharmaceutical human trials that are controlled and overseen by medical ethics boards and legislative bodies, commercial research, development, and testing of these vehicles currently occurs right alongside our other vehicles, pedestrians, and bicyclists. Mayoral consent, rather than strict governing bodies, is all that is currently needed to experiment with these systems.

The concept of replacement in society also elevates the question of employment writ large. Although there is no deep analysis of this topic in *Lo and Behold*, it is now becoming a significant part of societal discourse regarding AI's ability to, year by year, replace more of the jobs in which humans find dignity—for some, this is translated to a sense of self-worth—through careers and earned income. Fareed Zakaria, a *Washington Post* columnist, conducted an analysis of job categories across America and concluded that the single most common occupation in the United States is to drive a car, bus, or truck. His work suggests that driverless car technology converts human labor income into capital wealth owned by the corporate elite. This, in turn, exacerbates both underemployment and wealth inequity in society. Technology, so often advertised as a catalyst to create the human jobs that power the economic society, is beginning to show signs that it may be an existential threat to humans' abilities to derive income and productivity. Epochs long past demonstrated the value of a leisure class, but within the context of the United States the governmental infrastructure needed to support a workforce without work is not readily imaginable. Just the opposite is our current political climate, in which relief and universal support programs have dwindled to unprecedented levels of lowered support in many countries.

Life is no less strange than a science fiction story in this regard, and we need look no further than life itself to see how the newest AI and robotics technologies are engaged in rethinking the defining characteristics of society in a deep way. Entrepreneurial plays in the robotics space have taken up this mantle recently. This famously began with Cynthia Breazeal's Kickstarter campaign and company creation, Jibo, which was billed by its marketing arm as "the world's first social robot for the home."[17] But this project was rapidly eclipsed by heavyweight Amazon as it introduced the Echo, a more convenient and less expensive next iteration. The newest social robotics companies also pursue this line of thinking. Consider the 2018 marketing copy of Embodied, a stealth company that is making a new effort to introduce robots as social peers to humans: "Our team combines 14 years of research in socially assistive robotics from University of Southern California* with 15 years of experience with commercialization of consumer robotics from iRobot, and decades of animation and storytelling to build life-like, interactive robots to serve as motivators, coaches, and companions that help and empower people to be better at helping themselves."[18]

"Robots as companions" reconsiders altogether the station of machinery in human society. This is a prominent point of departure from the industrial designer's commitment to create well-behaved, predictable machines suitable for one specific task, such as toasting bread or heating food. In the eyes of even government regulation, this shift in technology from industrial to social has become so significant as to raise new, fundamental questions about the rights afforded to AI systems in the name of preserving the functionality of the legal liability and tort systems.

The European Parliament resolution with recommendations to the Commission on Civil Law Rules on Robotics was released in 2017 and explicitly lists for Commission consideration the creation of legal personhood for robots: "Creating a specific legal status for robots in the long run, so that at least the most sophisticated autonomous robots could be established as having the status of electronic persons responsible for making good any damage they may cause, and possibly applying electronic personality to cases where robots make autonomous decisions or otherwise interact with third parties independently."[19] This language suggests a depth of agency ascribed to certain robots, sufficient to have a robot be *responsible* itself for the damage it has wreaked and responsible for the remediation thereof.

The European Parliament resolution explains that machine learning represents the crux of the challenge. Because a robot or AI system can be put "in the wild" in such a manner that it learns over the course of its time in society, its actions thereafter are not attributable only to the engineer who created the system, but to its experiences over the course of its interactions with society through its consequential decision-making processes. The commission's solution, to consider robot personhood, suggests a concrete form of change to the definition of society and to the ways in which property, such as a robot, may have rights in view of interactions with society. This debate, and our collective witness of rapid technological changes that influence our understanding of society, are in their infancy. The subsequent chapters provide the frameworks and tools for you to begin to create your own conceptual framework for how this possible future may play out.

Discussion Questions

1. Suppose for a moment that we accept the premise that AI advances will enable the creation of machines that have ever-greater levels of apparent autonomy and agency. If a new *Keyword* entry for *society* were written twenty years hence, how might the meaning of *society* have changed because of such technological advances?

2. How would you propose that responsibility should be ascribed if a self-driving automobile causes harm?

3. As we consider the impact of technology on society, is AI any different from other technologies? If so, how is its impact unique or different?

4. *Class concept map:* Make a concept map in which you consider key concepts in relation to *Lo & Behold*. What relationships (arcs) connect concepts to one another? (See the sample assignments for a concept map example.)

3 Labor and the Self

Figure 3.1

Raw cotton is packed with seeds that make the fibre impossible to weave. In 1793, Eli Whitney patented a mechanical cotton gin that combed the seeds out of the soft fiber. Before Whitney's gin one person could clean one pound of cotton a day. The gin increased that number by 4,900%. This left a major bottleneck- picking enough cotton to fill the gins. Slave owners forced enslaved African Americans to work longer and harder and demanded more land in the west.
—Smithsonian National Museum of African American History and Culture

A civics lesson from a slaver. Hey neighbor.
Your debts are paid cuz you don't pay for labor
"We plant seeds in the South. We create." Yeah,
keep ranting
We know who's really doing the planting.
—Alexander Hamilton, in *Hamilton*[1]

Sources

Narrative of the Life of Frederick Douglass, an American Slave by Frederick Douglass

The Idea of the Self by Jerrold Seigel

RUR (Rossum's Universal Robots) by Karl Čapek

Keywords: Labor

Keywords for Today: Humanity

Guiding Issues

At the Smithsonian National Museum of African American History and Culture, curators engage with a public that brings varied relationships to the institution of American slavery. One challenge is to bring the scale of the institution, and its seeming distance in time, to life. Foregrounding the physicality of enslavement is a primary focus for exhibitions on the ground floor, where curators carefully arrange relics to demonstrate the constraints of enslavement on the individual man, woman, or child. Tools used for physical confinement, instruments for punishment, and machines used to automate key features of preparing cash crops for markets are documented and exhibited alongside a carefully paced and curated narrative. The tools range from child-sized chattels to chains to a cotton gin. The narrative of African enslavement is delivered through citations from Thomas Jefferson's writing, alongside testimonials from individual slaves, who articulate their lived experience in regard to legislative decisions that render them property, rather than persons or citizens, in a newly born republic. A tool like the cotton gin betrays technological advancement as a driving force, a catalyst. It demonstrates the propagation of an economic, social, and cultural system built on exploitation. Chattels on display illustrate one of several efforts, by slaveholders and legislators alike, to undermine individual agency, to compromise each slave's autonomy in buttressing an economic system that is positioned to relentlessly proliferate until legislative intervention, which did not come until the abolition of the slave trade in the early nineteenth century and the abolition of slavery decades thereafter.

In 1793, Whitney's gin not only replaced the need for an individual person to clean harvested cotton, it also optimized that effort. The cotton gin displaced human laborers, swapping in a system that was fifty times more efficient at this particular job function. It moved further labor demands from the cleaning of cotton back to the scaled growth and harvest of the crop. Today, language pertaining to automation echoes in mainstream media throughout the United States, decrying the prospect of replacing human labor with AI systems. But in Whitney's example, we know that replacing the human with a machine actually propagated a new demand for an even larger labor force needed to grow and harvest ever more cotton to "fill the gins." This became an economic pillar, foundational to the United States. As we consider the interface of artificial intelligence and humanity at present, how does dehumanization of a historically disenfranchised population suggest questions pertaining to dignity, self-determination, and citizenship in a new era of productivity? If labor might end as an individual's means to economic independence in the future, how do we navigate a new relationship between an individual, their state, and the economics of politics therein?

Language: Labor and Humanity

With the rise of capital in the eighteenth century in Britain and later, the United States, the interlinking of labor, and an individual's conception of humanity and self-determination became intertwined in interesting and dynamic ways. In *Keywords*, Williams writes:

> Labour was personified, as in Goldsmith's *The Traveller* (1764): "Nature … still grants her bliss at Labour's earnest call." But the most important change was the introduction of labour as a term in political economy: at first in an existing general sense, "the annual labour of every nation" (Adam Smith, *Wealth of Nations*, Intro.) but then as a measurable and calculable component: "Labour … is the real measure of the exchangeable value of all commodities." Where labour, in its most general use, had meant all productive work, it now came to mean that element of production which in combination with capital and materials produced commodities. This new specialized use belongs directly to the systematized understanding of capitalist productive relationships.[2]

Goldsmith's and Smith's respective descriptions, and the conceptual constructs signified in their language, connect to sovereign individual men (not women or children) and nations (not territories or colonies). Self-determination, freedom, and command over one's labor or a nation's "annual labor" are the necessary context and circumstance that allow for their respective configurations of *labor* (as a word in its contemporary sense) to make any sense.

In the case of the individuals who work to "fill" Whitney's gin, the notion of labor as the "real measure of the exchangeable value of all commodities" is skewed. This inherent flaw emerges because of the manner in which the labor force, bound in chattels of various sizes to accommodate child-to-adult enslavement, was in fact part of the capitalist property that combined "capital and materials" to produce "commodities," rather than laborers who contribute and participate in this system as individuals with agency and leverage to determine where their labor might place them in an economic system.

We know that in the United States and Britain alike, the horrors of working-class labor conditions also offered limited freedom. But in the shadow of slavery, these features of labor are certainly distinct. Presently, as we consider the prospect of individuals pushed out of current or near-future labor markets, what lessons might we learn from the links among senses of humanity, dignity, and self-determination that are so essential to participating in a society or economic system through one's labor? And how might we link some of these questions to versions of these issues that we have attended to in the past, with decidedly uneven results of success the world over?

Contemporaneous insights on conceptions of humanity offer an interesting counterpoint to the etymology of the keyword *labor* as well. In *Keywords for Today*, the complexity of the term *humanity* evolves in particular ways of note from the early sixteenth century onward. Echoes that link the concept of labor, around the eighteenth century, to features of humanity are recounted here:

> From the early sixteenth century, in English, the development is complex. The sense of courtesy and politeness is extended to kindness and generosity: "Humanitie … is a generall name to those vertues, in whome semeth to be a mutuall concorde and love, in the nature of man" (Elyot, 1531). But there is also, from the late fifteenth century, a use of humanity in distinction from *divinity*. This rested (cf. Panofsky) on the medieval substitution of a contrast between limited humanity and absolute *divinity* for the older classical contrast between humanity and that which was less than *human*, whether animal or (significantly) *barbaric*. From the sixteenth century there is then both controversy and complexity in the term, over a range from cultivated achievement to natural limitation. It was from this sense of some players as "neither having th' accent of Christians, nor the gait of Christian, pagan, nor man" that Shakespeare's Hamlet
>
> > thought of some of Natures Journey-men had made men, and not made them well, they imitated Humanity so abhominably. (*Hamlet*)
>
> But cf. "I would change my Humanity with a Baboone." (*Othello*).
>
> Yet the use of humanity to indicate, neutrally, a set of human characteristics or attributes is not really common, in its most abstract sense, before the eighteenth century, though thereafter it is very common indeed. There was the persistent sense of ranging from courtesy to kindness, and there was also the sense, developing from *umanità* and *humanitas,* of a particular kind of learning.[3]

The evolution of the term *humanity* in the eighteenth century, in the context of labor, might be interpreted by some as an encouraging beacon of an economic and political system based on equality. Its emergence in the period of rapid expansion of early capitalist markets, a nearing end to the slave trade but not the institution of slavery, and the rise of Enlightenment conceptions of the individual, person, and citizen are evident. As the evolution of the nation-state in the United Kingdom and the United States took shape, individuals historically prevented from shaping the political landscape were in fact participating at unprecedented rates. The evolution of the liberal state was in its infancy, yet individuals historically subject to the state were beginning to shape its features in the wave of eighteenth-century revolutions in France and the United States, leading to rudimentary democratic systems. Centuries still away from universal suffrage in the American and British contexts, the roots of these systems were emerging. Yet we know that participation and the power to rule in these systems still proves uneven at best today.

At the close of the eighteenth century, we have the contemporaneous emergence of the term *humanity* to indicate distinctions between the human individual and its counterpart in the "animal or … *barbaric.*" We have Goldsmith's equating nature with "Labour's earnest call," Smith's "annual labour of every nation," and Whitney's introduction of the cotton gin. If we are to configure these forces as just a few examples of a triangulation that includes economic development, labor force, and conceptions of the individual that come to shape features of personhood and citizenship, we have a rich foundation to consider past and present characteristics of anxieties and concerns pertaining to the economy, labor, and humanity.

Frederick Douglass

Leveraging his position from slave to person to citizen, Frederick Douglass addresses these threads in his abolitionist memoir, *Narrative of the Life of Frederick Douglass, an American Slave*. Recounting his education in basic reading and writing, he writes:

> During this time, my copy-book was the board fence, brick wall, and pavement; my pen and ink was a lump of chalk. With these, I learned mainly how to write. I then commenced and continued copying the Italics in Webster's Spelling Book, until I could make them all without looking on the book. By this time, my little Master Thomas had gone to school, and learned how to write, and had written over a number of copy-books. … When left thus, I used to spend the time in writing in the spaces left in Master Thomas's copy-book, copying what he had written. I continued to do this until I could write a hand very similar to that of Master Thomas. Thus, after a long, tedious effort for years, I finally succeeded in learning how to write.[4]

Douglass's initial education began with copying markings for the placement of boards in ship making: "A piece for the larboard side forward, would be marked thus—'L.F.' … I soon learned the names of these letters, and for what they were intended when placed upon a piece of timber in the ship-yard. I immediately commenced copying them, and in a short time was able to make the four letters named. After that, when I met with any boy who I knew could write, I would tell him I could write as well as he."[5] In this instance, Douglass's "copying" of the four letters, and his claim that "I could write as well as he" each time he meets any boy demonstrates a disruption to the power structure in which he lives. His emulation of other young boys who are expected to know how to write by a certain age indicates that Douglass assumes he too will participate as an actor in this society, not property that others will act upon. Douglass's education serves as a revolutionary example that configures relationships among developing economics, politics, technological advancement, and human power relationships from the late eighteenth century onward. As Douglass recounts graduating from copying the ship plank placement markings to a "lump of chalk" on the "board fence, brick wall and pavement," he is already initiating a significant recasting of his own relationship to the state in mid-nineteenth-century America. As a slave who learns to read and write, he initially mimics the master by copying the shapes of four letters in the shipyard. This evolves to learning their names and their significance. Thereafter he inhabits the political and juridical space between the slave and his "little Master Thomas," the young boy for whom he cares in Baltimore, Maryland. As Douglass spends his "time in writing in the spaces left in Master Thomas's copy-book, copying what he had written," he seizes this symbolic space in the copybook and in American society, usurping the rights that are categorically denied to him as a slave, including literacy, personhood, and citizenship.

In the copying of "little Master Thomas's copy-book," Douglass initially tests "the older classical contrast between humanity and that which was less than *human*, whether animal or (significantly) *barbaric*" that intellectually justified the system of slavery in the United States and Britain. Per Hamlet's quip: "O, there be players that I have seen play, and heard others praise, and that highly, not to speak it profanely, that, neither having the accent of Christians nor the gait of Christian, pagan, nor man, have so strutted and bellowed that I have thought some of Nature's journeymen had made men and not made them well, they imitated humanity so abominably" (*Hamlet*, act 3, scene 2).

Douglass's initial drawing of letter shapes and eventual copying of the master's lettering in the spaces literally left over in the blank spaces of the copybook demonstrate a seemingly blind mimicry. The features of a barbaric "gait of Christian" or "man" or perhaps those who "imitated Humanity so abhominably" seem familiar in the context of Douglass' mid-nineteenth century memoir. And yet, we know simply by virtue of reading the illuminating account that he captures in *Narrative of the Life of Frederick Douglass* that rhetorical genius, rather than beastlike mimicry, is actually at work instead.

As Douglass recounts being summoned back to Hillsborough to "be valued with the other property" pending their dispersal after Master Captain Anthony's death, this tension is palpable: "We were all ranked together at the valuation. Men and women, old and young, married and single, were ranked with horses, sheep, and swine. There were horses and men, cattle and women, pigs and children, all holding the same rank in the scale of being, and were all subjected to the same narrow examination."[6] In the etymology of *humanity*, the classical "contrast between humanity and that which was less than *human*, whether animal or (significantly) *barbaric*" is primarily couched in symbolic terms of language in which the signifier is linked to the signified philosophically. Douglass's recount of the "horses and men, cattle and women, pigs and children, all holding the same rank in the scale of being" suggests a visceral equation of being in the United States political context, in which slaves are property or capital, the very same as livestock. Without even an opportunity to exhibit mimicry of the master, mistress, or child but instead valuation beside livestock, Douglass's recount insists on a reconsideration of an enslavement system that strips the African population of its membership in "humanity." Literacy renders Douglass "unfit to be a slave," yet the political, economic, and cultural system in which he lives values him as property—not person, not citizen.

In *Keywords for Today*, we learn that "the use of humanity to indicate, neutrally, a set of human characteristics or attributes is not really common, in its most abstract sense, before the eighteenth century, though thereafter it is very common indeed. There was the persistent sense of ranging from courtesy to kindness, and there was also the sense, developing from *umanità* and *humanitas,* of a particular kind of learning."[7] As the turn of the eighteenth century saw a dramatic shift in describing a set of human features to distinguish humanity from animals or the barbaric populations, a "particular kind of learning" emerges significantly as well. An emphasis on agency demonstrated through literate means and verbal communication is valued as a distinct feature that separates the imitator of humanity from humanity.

Douglass emulates these distinctions in verbal and written utterance, narrating the plight of the enslaved African in a language and political register that is legible to his master. His comrades in the abolitionist movement also benefit from this significant shift in political and potential economic positioning by virtue of these changes:

> But, while attending an anti-slavery convention at Nantucket, on the 11th of August, 1841, I felt strongly moved to speak, and was at the same time much urged to do so by Mr. William C. Coffin, a gentleman who had heard me speak in the colored people's meeting at New Bedford. It was a severe cross, and I took it up reluctantly. The truth was, I felt myself a slave, and the idea of speaking to white people weighted me down. I spoke but a few moments, when I felt a degree of freedom, and said that I desired with considerable ease. From that time until now, I have been engaged in pleading the cause of my brethren—with what success, and with what devotion, I leave those acquainted with my labors to decide.[8]

As Douglass recounts speaking "but a few moments, when I felt a degree of freedom," he rhetorically marks the shift in his own psyche, a public display of this move, and the manner in which this movement indicates the prospect of a societal change. If the individual slave is rendered unfit to be a slave once literate, what are the social and political implications of the individual "pleading the cause of my brethren"? It is a revolutionary move that "those acquainted with [his] labors" now document and analyze as a monumental shift in American and British history. His memoir indicates one of several philosophical and rhetorical shifts in conceptions of the individual, the human, and the person that led several decades later to revolutionary shifts in political consciousness—and, thereafter, to changes in citizenship and universal suffrage. The public denouncement of the institution of slavery, however, was a fundamental first step, led in no small part by Douglass himself. These developments link in compelling ways to questions pertaining to personhood in machines through science fiction several decades later, as writers engage with technological, existential threats in the aftermath of the industrial revolution and the First World War.

Human Self, Cyborg Self

In *The Idea of the Self*, Jerrold Seigel documents the emergence of conceptions of individuality and self in Western philosophy, relying upon broad-stroke features of the corporal, reflective, and reflexive features of individuality as distinctions of humanity.[9] These categories of sentience are illustrated in high relief in Douglass's memoir and confirmed in compelling ways in the evolution of the word *humanity* in contemporary times. In *Keywords for Today*, MacCabe and Yanacek distinguish recent use of the term *humanity* as a linguistic indicator that separates sentient human beings from other living or artificial creatures:

> While the relative frequencies of human and humanity have remained fairly constant since the early nineteenth century, non-human (first attested in 1839 according to the *OED*) seems to have increased substantially in use in the mid twentieth century and again in the late twentieth century. This trend points to a growing interest in delineating the boundaries of humanity, that is, in distinguishing humans from machines and non-human animals and organisms. ... The noun and adjective post-human describe "a hypothetical species that might evolve from human beings, as by means of genetic or bionic augmentation." Post-humanism is used in two different senses: it can denote (i) a critique of the tenets of humanism and the idea of the autonomous, rational human subject, and (ii) "the idea that humanity can be transformed, transcended, or eliminated either by technological advances or the evolutionary process." No longer limited to science fiction, developments in robotics and artificial intelligence have made it possible to think of a post-human future in which humans will merge with software and machines in order to overcome biological limitations.[10]

Before moving to contemporary "developments in robotics and Artificial Intelligence" that make it possible to "think of a post-human future in which humans will merge with software machines in order to overcome biological limitations," it is useful to visit the imagined version of this future from 1920 in Karl Čapek's *(RUR) Rossum's Universal Robot*. Although we visit this work in Teresa Heffernan's forthcoming edited volume *Cyborg Futures* in depth, the significance of the play in dialogue with Douglass's memoir triangulates our analysis in combination with Douglass and Seigel.

In the 1920 farcical play, Čapek warns of a tech dystopia in which humanity is threatened by robotic systems that ultimately extinguish the human race. The posthuman qualities of the narrative demonstrate an exercise in exploring "the boundaries of humanity." Čapek moves from the Old Rossum's biological mimicry to the Young Rossum's engineering of humanoid laborers who outperform human labor forces without the complexity of biological mimicry. This transition reflects the manner in which conceptions of humanity (or perhaps European conceptions of humanity) are under threat when placed in direct contact (read opposition) with "the other," racial, ethnic, or in the science fiction iteration that Čapek introduces: a machine form of nonbiologic intelligence.[11] This concept of a European conception of individuality as not only distinct from but of higher value than other populations of humans deemed "other" and therefore deficient is clearly evident in Douglass's memoir and in postcolonial theory ranging from work by Edward Said to Franz Fanon, among others. The theatrical exploration in Čapek's play adds an explicit, farcical set of features. In *RUR*, the "other" is portrayed by machines, rather than another human population. They too are rendered deficient in comparison to their Western human counterparts. These oppositions are built, of course, on fundamental categories of subjugation that render the labor force a fleet of tools, not sentient beings. They are owned and operated by their human overseers. Čapek's vision suggests a category of beings distinct from "non-human animals and organisms," surpassing human labor forces in their utility and strength. They are merely a set of tools that do not threaten humanity until emotional capacities are introduced to their systems as a perceived technological enhancement, and thereafter as consequential decision-making is relinquished to these systems.

Soon after the play's debut in England, Čapek engages in public discourse pertaining to the play in the *Sunday Review*. He writes:

> The old inventor, Mr. Rossum, is no more or less than a typical representative of the scientific materialism of the last century. His desire to create an artificial man—in the chemical and biological, not the mechanical sense—is inspired by a foolish and obstinate wish to prove God unnecessary and meaningless. Young Rossum is the modern scientist, untroubled by metaphysical ideas; for him, scientific experiment is the road to industrial production; his is not concerned about proving, but rather manufacturing. To create a homunculus is a medieval idea; to bring it in line with the present century, this creation must be undertaken on the principle of mass production. We are in the grip of industrialism; the terrible machinery must

not stop, for if it does it would destroy the lives of thousands. It must, on the contrary, go on faster and faster, even though in the process it destroys thousands and thousands of lives. ... A product of the human brain has at last escaped from the control of human hands. This is the comedy of science.[12]

In his exploration of "the comedy of science," Čapek's scathing critique of the age of industrialization is politically motivated as well. Influenced heavily by the work of William James and others, Čapek was a strong critic of Soviet Union Communism and German National Socialism. As he espouses democratic principles, he criticizes ideologies based on scientific reason gone awry. Although not necessarily tech-averse, his farcical play suggests a need for rational development and humanist principles to safeguard humanity as a privileged category distinct from the synthetic, regardless of its acknowledged flaws. As a critic of oppressive systems, industrial, economic, or racialized, his work is also considered a thinly veiled critique of the legacies of slavery in North America and the Caribbean.

In *The Sound of Culture*, Louis Chude-Sokei suggests that Čapek's work attends to an imagined future wherein Western characters like Domin and the Rossums must face the implications of their ideologies brought to their rational conclusions. He writes, "The figure of the robot doubled and threatened the definition of white as exclusively rational while promising a figure of leisure similar to southern slave owners. *R.U.R.* after all, was about a robot revolution in which the machines not only rise up against their human masters and slaughter them but also evolve 'souls' of their own."[13] The self-suggested "savior" of the robots, presented through Helena, suggests an alternative to the "comedy of science." Yet her role also suggests the notion of the "white exclusively rational" in opposition to the robots as a tragic, misguided rationale as well. In the first act of the play, Helena demonstrates her allegiance to the perceived humanity of the robot race in her disputes with Domin, Alquist, Fabry, and Dr. Gall:

Domin: Dear Miss Glory, we've already had at least a hundred saviors and prophets here. Every boat brings another one. Missionaries, anarchists, the Salvation Army, everything imaginable. It would amaze you to know how many churches and lunatics there are in the world.

Helena: And you let them talk to the Robots?

Domin: Why not? So far they've all given up. The Robots remember everything, but nothing more. They don't even laugh at what people say. Actually, it's hard to believe. If it would interest you, dear Miss Glory, I'll take you to the Robot warehouse. There are about three hundred thousand of them there.

Busman: Three hundred forty-seven thousand.

Domin: Good. You can tell them whatever you want. You can read them the Bible, logarithms, or whatever you please. You can even preach to them about human rights.

Helena: Oh, I thought that … if someone were to show them a bit of love–

Fabry: Impossible, Miss Glory. Nothing is farther from being human than a Robot.

Helena: Why do you make them then?

Busman: Hahaha, that's a good one! Why do we make Robots!

Fabry: For work, Miss. One Robot can do the work of two and a half human laborers. The human machine, Miss Glory, was hopelessly imperfect. It needed to be done away with once and for all.

Busman: It was too costly.

Fabry: It was less than efficient. It couldn't keep up with modern technology. And secondly, it's great progress that … pardon.

Helena: What?

Fabry: Forgive me. It's great progress to give birth by machine. It's faster and more convenient. Any acceleration constitutes progress, Miss Glory. Nature had no grasp of the modern rate of work. From a technical standpoint the whole of childhood is pure nonsense. Simply wasted time. An untenable waste of time. And thirdly …

Helena: Oh, stop!

Fabry: I'm sorry. Let me ask you, what exactly does this League of—League of—League of Humanity of yours want?

Helena: We want first and foremost to protect the Robots and—and—and—to guarantee them—good treatment.

Fabry: That's not a bad goal. Machines should be treated well. Honestly, that makes me happy. I don't like damaged goods.[14]

Although Fabry insists that protecting his property aligns well with Helena's mission, he pauses upon learning that they "want to liberate the Robots." Helena takes the suggestion another step further when she states, "They should be treated like … treated like … like people." As the play unfolds, Helena's vision for liberating the robots does not materialize as she and her Humanity League activists imagine. The premise of just treatment, to ensure the labor force is treated "like people" as a means of liberation, certainly smacks of comedic undertones in a post–World War I context. This was a period in which the world witnessed technological tools that assisted with the obliteration of scores of "people" treated in a variety of inhumane ways, considered unprecedented in human history.

In the context of the industrial revolution throughout the United States and Europe, too, to be treated "like people" would not necessarily count as "good treatment" broadly conceived either. Chude-Sokei alludes to such circumstances as he writes in relation to links between the institutes of slavery and its ancestral heir in the early-twentieth-century technologically driven labor markets: "As a system, the plantation was a significant precursor to the regimentations and formal, time-driven deperson-alizations known as Fordism and Taylorism. It is this insight that motivated Carib-bean thinkers like the venerable C.L.R. James to stress the plantation as a 'dominant industrial structure' and to argue that it was on the plantation that slaves became disciplined into modern subjects in advance of formal freedom."[15] As we have triangu-lated the forces that come to define modern Western conceptions of humanity and its distinctions from its "non-human animals and organisms" or machines, we explore the trajectories of economic development, labor force, and conceptions of the individual that come to shape features of personhood. If Douglass breaks the chattels of slavery arguably through literacy, and Čapek's trope of the "comedy of science" is paired with the tragedy of "human rights activism" for the nonhuman machine, how might we configure our further exploration of links between artificial intelligence, humanity, and our evolving economic, political, and cultural systems as we move further into the twenty-first century? Chude-Sokei suggests that many of the conceptions and features of language that we might need to navigate the next pressing questions pertaining to personhood are evident in our past sins of exploitation and dehumanization.

Citing C. L. R. James, Chude-Sokei suggests that the spectrum of "depersonaliza-tion" is wide and historically long. It stretches from slavery and the plantation to Ford-ism and Taylorism. He writes that "blacks [are figured] as liminal, not-quite human beings in the age of racial slavery. They were or were like animals; they were or were like machines; and so they could be and were many things and were figured as such. It is that metaphoric flexibility—or hyperproductive lack as Sylvia Wynter might put it—that makes possible the long tradition of using blacks to either represent technology or metaphorically to oppose it; to use blacks as ciphers for machines or to use machines in ways that depend on earlier representations of blacks, even in futurism itself."[16] Čapek's fiction suggests a farcical thought exercise on the narrative trajectories of such "lim-inal, not-quite human beings," evident in a collective Western history. It is a narrative captured in poignant ways by the Smithsonian National Museum of African American History and Culture's portrayal of the influence of a tool like the cotton gin on human-ity. As we make our way into the next chapters of human history, we can interrogate the likelihood that *posthumanism*—"used in two different senses: it can denote (i) a critique of the tenets of humanism and the idea of the autonomous, rational human subject, and (ii) 'the idea that humanity can be transformed, transcended, or eliminated either by technological advances or the evolutionary process' "—may or may not signify a keyword in our collective and evolving history.[17]

Discussion Questions

1. Consider the application of the concept of *less than human* to a new machine feature: the intelligent machinery of AI. Is it justifiable to label such machinery as less than human, or does this judgement fail to apply because AI is too orthogonal to animal and human form to be comparable?
2. The ascription of personhood granted through the reality of labor is one angle for considering anthropomorphic qualities applied to intelligent machines. What considerations regarding labor performed by a machine would convincingly open up the ascription of personhood?
3. Helena argues for *robot rights* in *RUR*, before her plans fall apart. What are the conditions under which robot rights should be afforded? Why is this socially valuable under such conditions?

4 (In)equality and (Post)humanity

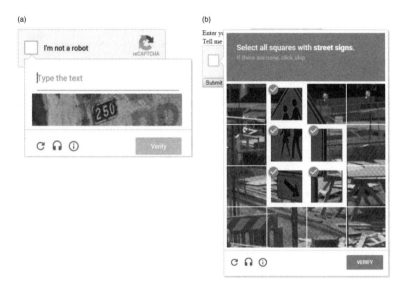

Figure 4.1a, b
reCAPTCHA humanity verification screenshots, circa 2017.

The Singularity will represent the culmination of the merger of our biological thinking and existence with our technology, resulting in a world that is still human but that transcends our biological roots. There will be no distinction, post-Singularity, between human and machine or between physical and virtual reality. If you wonder what will remain unequivocally human in such a world, it's simply this quality: ours is the species that inherently seeks to extend its physical and mental reach beyond current limitations.[1]

—Raymond Kurzweil, *The Singularity Is Near*

The "criticism from the rich-poor divide": *It's likely that through these technologies the rich may obtain certain opportunities that the rest of humankind does not have access to.* This, of course, would be nothing new, but I would point out that because of the ongoing exponential growth of price-performance, all of these technologies quickly become so inexpensive to become almost free.[2]

—Raymond Kurzweil, *The Singularity Is Near*

Sources

Keywords for Today: Equality, Humanity

Simon Head, *Mindless*, 1–13, 29–49, and 103–127

Guiding Issues

In the previous chapter, we studied how cases of human labor and machine labor have both led to usurpation of individual power, both in the case of historical reality, as in slavery, and in the imagined possible future of Čapek's *RUR*. Now we take our analysis of labor in hand and revisit history once again. This time we examine how the four industrial revolutions provide a chronological context into the interrelationships of labor, humanity, and equality. The terms *equality* and *humanity* provide this chapter's backbone, particularly because their definitions hinge on the dynamics of their negatives, *unequal*, *nonhuman*, and *post-human*, which sharply interrogate our considerations of the present and future impact of AI on society.

We consider the interplay of *equality* and *humanity* chronologically, visiting each of the four commonly recognized industrial revolutions. Each epoch of technological transformation provides us with new examples of how technical innovation leads to changes in labor conditions, in dignity for a labor force, and in the achievement of unequal stations by citizens. Two authors provide us with ethnographic, fine-grained observations that color the analyses of this chapter: Barbara Ehrenreich experienced low-income labor in America firsthand, living in poverty conditions and working as a waitress, hotel maid, house cleaner, nursing home attendant, and Walmart salesperson, as described in *Nickel and Dimed*.[3] And Simon Head conducts ethnographic research observation of working conditions in Walmart and Amazon, concentrating on how technology has influenced worker agency and empowerment in low-wage conditions in his book *Mindless*.[4]

Finally, after observing the influence of technology on labor historically and in the present moment, we turn our attention toward the near future, studying ways in which digital labor and extreme forms of AI and robotic surveillance have the potential to further alter equality and humanity in the present, fourth industrial revolution. The interplay between human agency and AI-based forms of digital exploitation will enable us to open a discussion concerning the blurring boundary between AI and humanity and the ways in which this indistinct middle space gives rise to societal disruption along the twin avenues of posthuman identity and technological inequality.

Language: Equality and Humanity

Two keywords, *equality* and *humanity*, form the basis of analysis for this chapter, which focuses on the ways in which technology has historically influenced human station and human identity. In *Keywords for Today* entries for both terms, you will note the striking use of each term's negative(s) as a critical tool for understanding the terms themselves. Although *equality* has been in use, as noted, since the fifteenth century, the use of the term most relevant to our analysis is coupled to the American and French revolutions in the eighteenth century.

Keywords for Today points out two branches of use for *equality*, one regarding a process of equalization and the second a starting condition to abolish unfair privilege. In both cases, whether equality is a goal or a starting criterion, inequality becomes the key component driving the dynamics of change. In the first case, inequality is the starting state, and the goal of equality is to reduce that inequality for normative, even ethical, purposes: "(i) a process of equalization, from the fundamental premise that all men are naturally equal as human beings, though not at all necessarily in particular attributes. ... (i) a process of continual equalization, in which any condition, inherited or newly created which sets some men above others or gives them power over others, has to be removed or diminished in the name of the normative principle (which, as in Milton's use, brings equality and *fraternity* very close in meaning)."[5]

In the second case, inequality is a station that can be arrived at *fairly* by ensuring equality of opportunity at the starting gate: "(ii) a process of removal of inherent privileges, from the premise that all men should 'start equal,' though the purpose or effect of this may then be that they become unequal in achievement or condition. ... (ii) a process of abolishing or diminishing *privileges*, in which the moral notion of equality is on the whole limited to initial conditions, any subsequent inequalities being seen as either inevitable or right. The most common form of sense (ii) is equality of opportunity, which can be glossed as 'equal opportunity to become unequal.'"[6]

Before we consider how AI influences these two modern branches of equality, it is worth considering how earlier automation and digital technologies have historically influenced equality. Turning the question on its back, we can restate the issue: How has technology historically affected how we are *unequal*? The most classical economic response stems from agricultural technology, summarizing that most of the land's population used to be dedicated to the concern of growing food. Agricultural technology inverted that number so that now less than 3 percent of the population is dedicated to this task today. Global quality of life, diversity of careers, and indeed city planning itself are all results of this massive change in human agricultural productivity per capita.

Certainly, we can attribute dramatic improvements in quality of life not only to agricultural technology, but also to medical technology and many other forms of innovation. Yet if we return to the question of being unequal, we face an interesting puzzle: Subpopulations of those who are technologically fluent benefit especially from technology innovation. But technology particularly privileges those who are fluent. If technology fluency were considered a characteristic like socioeconomic rank or gender, then we would see it as a privilege. Its discriminating power would, in turn, counter equality. Indeed, *Keywords for Today* notes that any measurable difference that gives one group power over another threatens at least the first branch of equality. If, alternatively, technology fluency is a discriminatory power that is the endgame of that "opportunity to become unequal," then it is the precise mechanism by which technological innovation might make us more unequal over time.

Even as you consider the role of technology through inequality, suppose further that artificial intelligence advances on its current track to a point at which it moves well beyond augmenting a human user and instead stands in the position of replacing a person for any given task. If AI is as highly functional as a person, then might it attain a station of privilege above that of certain subclasses of people? Can an AI be unequal to a group of people and by this mechanism force a loss of equality among persons? This edge case of equality, in which a thinking and acting machine introduces a new dynamic of the unequal into our prior understandings of social equality, is a second open question worth exploring.

A future-facing analysis of equality thus begs an understanding of unequal, even between human and machine. This naturally leads to a return to the keyword *humanity*. Under "Recent Developments," *Keywords for Today* notes trends toward the use of the word *nonhuman* and the even more recent use of *posthuman*. Once again, the definition is influenced strongly by its negative uses, just as with *equality*:

> While the relative frequencies of human and humanity have remained fairly constant since the early nineteenth century, non-human (first attested in 1839 according to the *OED*) seems to have increased substantially in use in the mid twentieth century and again in the late twentieth century. This trend points to a growing interest in delineating the boundaries of humanity, that is, in distinguishing humans from machines and non-human animals and organisms. … Important twentieth-century coinages first appearing in the third edition of the *OED* include post-human and post-humanism. The noun and adjective post-human describe "a hypothetical species that might evolve from human beings, as by means of genetic or bionic augmentation." Post-humanism is used in two different senses: it can denote (i) a critique of the tenets of humanism and the idea of the autonomous, rational human subject, and (ii) "the idea that humanity can be transformed, transcended, or eliminated either by technological advances or the evolutionary process."[7]

These usage trends, particularly in the case of *non*human and *post*human, demonstrate that we have reached a boundary case in society. Our biological, deterministic dissection of the human body started a deconstruction process that makes the human, to a natural scientist, seem less unique, less magical compared to its constituent parts or to animals: "In her recent book *10% Human* (2015), biologist Alanna Collen discusses how humans consist of more colonies of microbes than human cells: 'You are just 10% human. For every one of the cells that make up the vessel that you call your body, there are nine imposter cells hitching a ride.'"[8]

Deconstructionist views of the human within, questioning the essence of humanity itself, is just a first, intrinsic part of a larger trend transgressing biologic boundaries. As machines begin to emulate aspects of human behavior, and even overperform spectacularly on the human cognitive scale (Go, chess, radiological exams, patent law, contract analysis, etc.), our apparent human uniqueness becomes further diminished. By the twentieth century, the idea of a species that is an evolutionary descendant of the human takes hold with the word *posthuman*. *Keywords for Today* specifically calls out robotics and artificial intelligence as potential candidates for overcoming (read: moving beyond) human form. *Keywords* closes the *humanity* entry, aptly, with a famed Stephen Hawking quote that ironically suggests that the most important factors regarding the future of humanity likely derive from the nonhuman machine: "Optimism about the potential of technology to improve human life has been matched with skepticism and fear that it could threaten the future of humanity, as Stephen Hawking warned: 'The rise of powerful AI will either be the best or the worst thing ever to happen to humanity. We do not yet know which.'"[9]

Four Industrial Revolutions

A review of the three prior industrial revolutions will provide context and understanding for making sense of the present fourth industrial revolution, also called the second machine age.[10] We wish to focus on the ways in which all four industrial revolutions have directly impacted the relationship between labor and capital, which is a critical dynamic in understanding the nature of labor, the ways in which inequality has evolved, and the ways in which changes in labor conditions impact human dignity and thereby the identity of humanity.

Technological automation has redefined labor in two distinct ways: capital conversion and labor transformation. Both mechanisms have significantly altered wealth disparities and working conditions, directly impacting inequality and humanity. Consider the first industrial revolution, which commonly refers to eighteenth- and nineteenth-century European and American mechanical innovations in textiles and agriculture. Industrial-age machines, from the cotton gin to grinding and sorting systems for wheat, were driven by rotary power supplied, disruptively, by a new invention: the steam engine. This period, commonly referred to as the Industrial Revolution, marks the sudden increases in local production efficiency, but productivity led to countless unintended side effects: dangerous, dehumanizing working conditions in new factories; displacement of large groups of workers as their jobs were replaced by machinery; blossoming of industrialist wealth; and significant increases in the standard of living for the wealthy class. As described in the previous chapter, automation and its related productivity increases in one part of a supply chain, as witnessed by the introduction of the cotton gin, can directly impact labor pressure and inequality in the rest of the supply chain, such as impacts due to the massive increased demand for raw, picked cotton. Increased productivity, even in isolation, is not without its consequences.

Furthermore, job displacement, commonly derided as a short-term hiccup by modern technology optimists, is often dismissed by pointing out the Luddite movement of the time. A frequently cited dynamic is that new jobs were invented because the Industrial Revolution created whole new categories of products and services. Although true, it is also important to note that *short-term* is a relative term. In Scotland, for instance, massive layoffs in textiles led to general, abject poverty for two solid generations before a recovery ensued, leading many Scots to leave, in desperation, for the shores of the United States.[11]

The first industrial revolution resulted in wealth generation through the first labor-changing pathway, *capital conversion*, which captures the logic of return on investment in capital investments. As machinery is invented that is able to do even a fraction of the work of human laborers, then the business owner must interrogate the calculus of investment and returns: Will investment in the machinery either decrease labor costs or increase revenues, and will it do so quickly enough to recoup the cost of the automation equipment? If indeed the machine can pay for itself by reducing the need for skilled human labor, then its acquisition converts work done by human labor, which previously achieved *labor income* for workers, into work done by capital equipment owned by the corporation, resulting in *capital income* for the owner. So automation can stand as a pump from labor income to capital income, changing not only how work occurs, but also how compensation for work is distributed in society. The example of automated textile machinery in the early 1800s England, during the first industrial revolution, represents such a labor-to-capital pump, with automated looms replacing the work of human laborers and promising returns far exceeding the cost of the automated machinery to the mill owners.[12]

The *second industrial revolution* refers to the creation of electric power production and, with it, the invention of machinery that uses electricity rather than steam power to operate. This power innovation massively increased the geographic footprint of industrial activity thanks to the ability to lay electric wire rapidly, and the electric light disruptively changed laboring hours and nighttime working conditions. The internal combustion engine and the extraction of oil together led to a universal, low-cost, and lightweight way of providing power anywhere, improving vastly on the heavy, expensive, and large steam engine as the power plant of choice.

The second labor dynamic, *labor transformation*, plays a large role in the second industrial revolution thanks to the invention of the assembly line. Consider the famous early assembly line example, the Ford Model T motor car. Henry Ford famously discovered that instead of having each worker concentrate on a single automobile from start to finish, he could have each worker do just one, tiny job repeatedly for every automobile chassis rolling through the line. Thus the work of laborers, given the right technology, becomes hyperspecialized in a way that increases whole-system productivity greatly. Ironically, given the labor-to-capital pump that all corporate owners wish for, the labor transformation dynamic is one in which the work left to humans is pointedly that work that automation cannot yet displace affordably, and the technological innovation tends to reduce human work to this most essential grammar of human action. The reductive nature of such reprogramming, though greatly increasing productivity, is not designed to serve the laborer's quality of life in any way. Repetitive stress, whether in the original Model T line or in today's Chinese megafactories, arises because a human performing the same small task repetitively is subject to injuries born of the unnaturalness and indignity of such work.

The third industrial revolution, the digital revolution, represents the primacy of the digital computer and all its associated products, systems, and workplace innovations. The minicomputer changed how companies perform complex calculations. The personal computer changed how middle-class homeowners store and manipulate data, and then the internet disruptively shifted how we communicate with one another and how we discover information and conduct commerce, seek and perform employment, and so on. The digital revolution is frequently used as evidence in debates regarding current job dynamics because the modern computer certainly displaced certain job categories (secretaries, accountants, etc.), but arguably created a vast new class of computer-based jobs throughout society. It is a legible success story frequently held up as an extrapolated "proof" that the present changes in the age of AI will surely be for the best as well.

The real impact of the digital revolution was neither simply job displacement nor job creation. Its greatest disruption centered on *information*, in how digital technologies

have changed power relationships in the workplace and between consumers and corporations. In *Mindless*, Simon Head shows how relatively primitive digital surveillance
and tracking systems change the nature of agency in labor by creating a system of
observation, analysis, and mathematical optimization that takes control of a laborer's
fine-grained routines. He writes of the Task Manager program at Walmart:

> Foremost among these technologies is "Task Manager," a targeting and monitoring system that
> Walmart began to introduce in its stores from 2010 onward. The system tells employees what
> to do, how long they have to do it, and whether they have met their target times. Employees
> sign on to the system by swiping their identity cards on a terminal, as with a credit card, and
> the system then spits out its instruction. In a research paper on Walmart's "Productivity Loop,"
> John Marshall of the Capital Stewardships Program of the Union of Food and Commercial
> Workers (UFCW) found that Task Manager is "an object of scorn among thousands of Walmart
> associates" who complain that "there is never enough time to complete all the tasks."[13]

Emphasizing worker productivity is nothing new, but digital innovations such as card-
based, real-time hour tracking change the resolution of surveillance. They introduce
new levels of quantification, oversight, and prescriptive control over ever-finer levels of
detail in the worker's day-to-day behavior. Studying Amazon, Head further notes that
electronic position tracking of warehouse employees enables managers to see how the
workers move throughout the building. If a worker visits a toilet that is not optimal,
the manager calls the employee out on that nonoptimal behavior: "The functional
foreman would record how often the packers went to the bathroom and, if they had
not gone to the bathroom nearest the line, why not."[14]

Surveillance of workers also is nothing new. But it is worth considering the ways in
which digital technologies have tipped the scale further to the advantage of business
owners and away from workers. The threat of digital surveillance, even unrealized, can
provide a strong power dynamic over workers. Barbara Ehrenreich reports on the effect
of this during her time as a house cleaner:

> When I ask a teammate why the rule against cursing in houses, she says that owners have
> been known to leave tape recorders going while we work. Video cameras are another part of
> the lore, positioned near valuables to catch a cleaner in an act of theft. Whether any of this is
> true or not, Ted encourages us to imagine that we are under surveillance at all times in each
> house. Other owners set traps for us. In one house, I am reprimanded by the team leader for
> failing to vacuum far enough under the Persian rugs scattered around on the hardwood floors,
> because this owner likes to leave little mounds of dirt there just so she can see if they're still
> there when we're done.[15]

As cameras and networks become less expensive globally, we face the prospect of pervasive surveillance, frequently generating data owned by power entities, whether corporate or government. This increasing degree of information asymmetry in the hands

of hegemonic power structures can further erode the feeling of personal agency and privacy felt by individuals throughout society. Surveillance viewed thus can be a blunt instrument that further exacerbates the unequal.

The *fourth industrial revolution*, a moniker initially coined by the World Economic Forum to refer to the newest innovations in labor and productivity, which can include near-future robotic and AI innovation, refers to an age when digital technology ceases to be trapped within the confines of a desktop computer. The Internet of Things promises the insertion of computational technology in virtually all physical products. The *quantified self* suggests the digital capture of every human sense and action, from heart rate and blush response to emotional fluency and activity tracking. Digital and physical surveillance unite to provide a complete picture of each person's physical and online activities in a unified holistic picture. Massive networking promises that information is richly fused, creating secondary knowledge that was otherwise impossible to capture, such as demographic data, purchasing habits, and the chances of loan defaults. Robotic innovations threaten prevailing concepts of aging and injury, suggesting exoskeletons and other robotic orthotics that change how humans and machines can couple in the physical world.

The list is long, and the critical notion is that the fourth industrial revolution gives digital technology the chance to break out of the computer and pervade the entire world, redefining our holistic relationships to computation and digital intelligence, as well as redefining our human relationship to the corporations that surround us. Discourse regarding technology that replaces human cognitive labor is not at all new. Norbert Wiener, who famously coined the term *cybernetics* in the 1940s, was already concerned about the power relationship and dynamics of automation and human labor by 1950. Yasha Levine[16] cites Wiener's work in *The Human Use of Human Beings: Cybernetics and Society*: "Let us remember that the automatic machine, whatever we think of any feelings it may have or may not have, is the precise economic equivalent of slave labor. Any labor which competes with slave labor must accept the economic conditions of slave labor. It is perfectly clear that this will produce an unemployment situation, in comparison with which the present recession and even the depression of the thirties will seem a pleasant joke."[17] Wiener's words are stunning when you consider the state of computational technology in 1950—nearly seventy years ago. He was already concerned with a level of computer sophistication beyond even the dreams of most engineers—and during the second industrial revolution.

Fast-forward to today's fourth industrial revolution, and in terms of human skill, the present innovation trend is toward capital equipment that can replace human action well beyond just manual labor and ever further into the regime of cognitive human functions. Today, law firms that do not have an AI contracts reviewer are seen as old-school and disadvantaged. Dominant practice suggests every law firm ought to be use an AI-based clerk to scour contracts for signs of ill intent and poor contract wording.[18] Many Fortune 50 companies refuse to pay for hourly work done by first-year legal clerks based on the presumption that all such work ought to be performed using cheaper, more "sophisticated" AI software. Automated sports page journalists, such as the *Washington Post*'s Heliograf, commonly employed to write a summary of yesterday's sporting news, have replaced an entry-level human journalism job wholly, providing articles that are not necessarily better but are considered good enough.[19] Even interactive tax-filing systems, such as TurboTax, provide greatly reduced tax-filing costs to individual citizens, but at the expense of replacing the work of an entire class of entry-level tax accountants. In this final example, the labor-to-capital pump is very direct indeed: the labor income of an army of tax accountants has been converted into capital-derived revenues for Intuit.

As we consider the near-future pathway of technological progress and its impact on this labor-to-capital pump, we can be confident about one trend: the frontier of computational work that enables a profitable labor-to-capital pump for corporations is always advancing further. Researchers in machine learning and artificial intelligence move the boundary in one direction: toward the advantage of cheaper, more capable computers that can do more of what we often consider the province of human thought and action. Furthermore, motor and sensor technology continues to improve the physiological capabilities of robots and computers, putting ever more sophisticated mechanical systems within grasp of robotic control. Sophisticated walking and running, hand-based grasping, and touch-based manipulation—all these landmarks of physical motion are today being overcome by researchers at decreasing robot price points and increasing capability. No single job, from a labor-to-capital pump perspective, is safe over the long term from machine automation.

If we consider labor transformation rather than labor replacement, AI has the potential to continue the trend of labor "deconstruction." Specifically, artificial intelligence enables corporate owners to further investigate, model, and reduce the minutiae of human labor into a corner case in which it happens to be more efficient and more productive than present-day automation equipment. Modern vision systems can accurately detect humans and their trajectory of motion and quickly convey this information to an informational dashboard. Such systems often depend upon specialized models of the human musculoskeletal system, enabling calculation of human efficiency of motion.

Head reports on the ways in which micromanagement of human behavior helped Amazon institute previously unjustifiable quota requirements of its warehouse workers.

Studying the time-motion behavior of workers, Amazon's algorithms recognized that employees were not using the most time-optimal possible gaits and coordinated moves to lift boxes from shelves. By creating training centered on new ways to cross legs, simultaneously reach out with one's hands, and then bounce into a reverse motion, Amazon justified an expectation of several tenths of a second savings on each and every box retrieval in the warehouse, upping quota requirements as a result.

The irony of such an approach to profit maximization is the extreme power dynamic constituted by micromanaging a human being's muscular motions: the warehouse workers are being *remote-controlled* by computer, as if they are the hands and legs of a robotic system. The only reason they are still human, in this case, is that they still cost less overall than their eventual machine replacements.

Both Simon Head and Barbara Ehrenreich report on *time theft*, a form of micro-managed power expressed by Walmart over its employees when they veer away from optimal behavior. Ehrenreich reports on the use of this moniker during initial new employee training: "The old guy who is being hired as a people greeter wants to know, 'What is time theft?' Answer: Doing anything other than working during company time, anything at all."[20]

New technology takes this trope to a whole new level. Head reports on how position-tracking of employees enables Amazon to literally count the total number of seconds of hesitation during movement—for instance, when an employee "illegally" checks their mobile phone and pauses in a corridor. Today's surveillance systems enable the corporation to sum up the total number of minutes "stolen" from Amazon, and this precise quantification empowers the company to back-charge the employee for stolen time. Technology does not invent abuses of power, but it constitutes increasingly brilliant tools for exacerbating the unequal relationships between owners and employees, leading to further employee indignity.

Head further notes how the emotionally destructive power dynamic of surveillance and information asymmetry erodes the effectiveness with which employees can do their jobs, and even relate positively to the corporation's customers:

> The multifaceted deterioration in the condition of labor helps explain why the theorists and practitioners of emotional labor devote so much effort to the repression of negative, work-disrupting emotions before trying to replace them with more customer-friendly ones, which they think will be good for sales. But looking at this project with even a modicum of historical perspective, it seems doomed. The harsh, unforgiving workplace described by Kochan yields a negative emotional labor all its own, and the longer this workplace endures, the more entrenched become the emotional armor that employees must rely on to protect themselves against a hostile world, thus cynicism, resentment, emotional withdrawal, and the withholding of loyalty from employers who show no loyalty to them.[21]

Even as we create AI systems that challenge our notions of the intelligence of a nonhuman entity and reflectively interrogate our understanding of what constitutes human form, the same machine learning and computational technologies have the power to create working conditions for human laborers that are less humane, leading to an ironic, symmetrical closing of the gap between the humanity of machines and the machine-like treatment of people.

Recognition of this changing worker-corporation dynamic is essential to making sense of the ways in which AI might change society. In common discourse, the standard concern aired regards unemployment and automation: Will automation reduce the size of the job market? Or will innovation lead to new categories of jobs at a rate exceeding this labor-conversion effect? We ask the wrong question: in the year 2019, we live in a US society with the lowest *published* unemployment rate in nineteen years. However, unemployment counts only those who continue to participate in the attempt to find jobs. The *labor force participation rate*, by contrast, has rarely been lower than it is now since World War II.[22] In very localized ways, the specter of machine autonomy has resulted in very odd participation rate dynamics: long-haul truck drivers are retiring rapidly, creating the biggest demand for new long-haul driver hiring in many decades. Yet there are few applicants because those entering the labor market see the writing on the walls due to the self-driving truck trope that infuses national media. So we have both the possibility of widespread displacement of long-haul drivers and a massive understaffing of open long-haul driver positions in the interim.

The machine age has a complex, textured impact on the human laborer. More than anything, it has changed the nature of labor. Ehrenreich eloquently makes the case that we fail to understand that employment itself is no longer a sufficient goal in society. Full-time work correlates, for a large class of low-income workers, with full-time poverty:

> When unemployment causes poverty, we know how to state the problem—typically, "the economy isn't growing fast enough"—and we know what the traditional liberal solution is—"full employment." But when we have full or nearly full employment, when jobs are available to any job seeker who can get to them, then the problem goes deeper and begins to cut into that web of expectations that make up the "social contract." According to a recent poll conducted by Jobs for the Future, a Boston-based employment research firm, 94 percent of Americans agree that "people who work fulltime should be able to earn enough to keep their families out of poverty." I grew up hearing over and over, to the point of tedium, that "hard work" was the secret of success: "Work hard and you'll get ahead" or "It's hard work that got us where we are." No one ever said that you could work hard—harder even than you ever thought possible—and still find yourself sinking ever deeper into poverty and debt. When poor single mothers had the option of remaining out of the labor force on welfare, the middle and upper class tended to view them with a certain impatience, if not disgust. ... But now that

the government has largely withdrawn its "handouts," now that the overwhelming majority of the poor are out there toiling in Walmart or Wendy's—well, what are we to think of them?[23]

Furthermore, unemployment itself has become a false proxy for the state of employment in the United States. As described in *Men without Work* by Eberstadt, even as we have entered a time of record unemployment, the total number of prime-aged men who are not active in the labor force has climbed explosively, from seven hundred thousand in the late 1960s to more than seven million today. Thanks to the efficiency of the overall economic system, high productivity and GDP growth continue unabated even as massive subgroups of our citizens lose the will to even look for employment, falling out of the statistics of the unemployment rate altogether.

If technology plays a deleterious role in this dynamic, that role includes the element of the *unequal*. Surveillance, job transformation, worker optimization: all these digitally inspired corporate improvements exacerbate trends that are already baked into our corporate handbook: the disempowerment of low-income employees and the further erosion of their agency and dignity in labor. The trend lines we witness are not cause for optimism. AI-based corporate innovation does not appear to be subject to natural corrective forces that would lead to greater equality and greater recognized humanity for all workers. This lack of a well-engineered, productive roadmap is both cause for alarm and a justification for considering how the fourth industrial revolution should be actively managed for humanity rather than posthumanity.

The Consumer as Digital Laborer

The prior section discussed the dynamics of corporations and employees through the lens of the four industrial revolutions. Technology also has a distinct influence on another relationship: that of the consumer and the corporation. *Digital labor* is a modern term describing ways in which corporations have begun to disruptively monetize consumer behavior, creating corporate income out of raw consumer action at a previously unprecedented level.[24]

The present concern is how and whether this newly programmed relationship between consumers and businesses constitutes a new form of labor exploitation. Exploitation implicitly requires a relationship between a subaltern group's labor and the misdirection of profits to a dominant class. Slavery is a classical example that we draw upon to consider both exploitation itself and the role of narrative in making sense of the history of labor exploitation. To consider the role of historical technology and future-facing AI in such imbalances of power, it is worthwhile to consider a three-way concept map among the subaltern, the dominant, and technology innovation.

We previously discussed the role of technology in enabling imbalances of power to resist revolt. The transatlantic slave trade required oceangoing vessels, which in turn depended upon new forms of celestial navigation, forecasting technology, and voyage-planning strategies. The very ability of a small crew, in ship and on land, to establish hegemonic dominance over a massive class of slaves demanded technologies for assertion of power: locks to restrict motion, devices for punishment and corporal harm. Technology imbues the owner with an advantage that tips balances of power, and with the forward evolution of technology, it does so with greater ratios of power and over larger distances. Literacy itself, as eloquently demonstrated by Douglass, was driven by technological advancement, redistributing negotiations of political and societal power to undercut the institution of slavery, in addition to other revolutionary disruptions.

We have also previously discussed modern technology in relation to entrenchment of power. Surveillance technology crystallizes information asymmetry, a critical currency for negotiations of power, to the surveillance ownership class. As described by Head, when Amazon is able to track the motion of every warehouse stocking worker and detect when each person slows down in an aisle to check their text messages instead of moving at full walking speed, then the corporation has gained an unchecked new capability to literally calculate the monetary opportunity cost derived from the most subtle of human behaviors—that of slowing and attending to a personal matter. When this act is quantified, proven on video, and packaged for sharing, it is reified to *time theft*—and is literally available as a justification for employee chargeback.

Technology does not itself invent such power relationships; the floor manager has always had this power. But the technological surveillance available today is newly omniscient, and this changes the power dynamic considerably. This basic notion—that technology frequently does not invent a new power dynamic, but greatly decreases friction, leading to a level of optimization heretofore unimagined—is a recurring trope. It is an argument often used by defenders of innovation to explain that *technology is neutral*. The argument suggests that *because technology simply implements existing processes, it cannot be unethical*. However, this argument ignores the fact that even questions of degree are material to the human condition, just as a bomb that kills many is freighted with ethics well beyond those of a bullet that kills one.

Turning now to the consumer-corporation relationship once again, note that, classically, consumers have been seen as rational economic agents: corporations provide marketing information, consumers make purchasing decisions, and money flows to the "winning" companies. Monetization at the commercial level rests with convincing consumers to give corporations their money and therein their purchasing loyalty.

In the past two decades, the big data revolution has suggested that given sufficiently massive data about consumer behavior, companies have the ability to model future consumer behavior at a level of accuracy that was previously unimaginable. This power stems from the application of machine learning to big data. The computer can find patterns that are highly individualized in massive datasets and can use these patterns to learn how to predict future behavior better than any human being. As the urban legend puts it, advertise beer next to diapers in the supermarket and the computer can very accurately predict exactly which demographic will purchase the beer when out on a diaper run. If the computer cannot make these predictions, then you simply do not have enough data; bring in more data and, eventually, predictions will accurately reflect reality.

This concept of technological optimism relating to data-rich behavioral analytics has risen with such force and conviction that it has become *frothy*, to borrow a term from Alan Greenspan. Valuations of companies ride on the amount of data they can collect from consumers, not because they have effectively shown monetization from the data but because of the mere fact that the venture capital community believes that massive amounts of data must contain gemstones of potential value, waiting to be mined eventually by clever AI systems in the near future.

Even companies that sell physical products to make profit are forced by their boards and investors to reconsider their underlying motives and to collect as much data as possible from consumers. Supermarkets no longer make all their money selling their produce and manufactured goods. They give you loyalty cards with which they track your purchasing behaviors precisely. Then supermarkets sell this purchasing behavior to marketing analytics companies. The marketing analytics companies perform machine learning procedures, slicing the data in new ways, and resell behavioral data back to product manufacturers as marketing insights. When data and machine learning become currencies of value in a capitalist system, then every company's natural tendency becomes to maximize its ability to conduct surveillance on its own customers because the customers are themselves the new value-creation devices.

The concept of customer-produced value is called *digital labor*, and this rapidly accelerating trend creates billions of dollars of value for corporations.[25] It is worthwhile to ask: Where does this value come from? That is, who is paying the price for these increasing valuations? The answer, put simply, is that digital labor is taxing the citizen. It is a tax on the time of every person, and taken all together that tax accumulates into consequential value for corporations. Two oft-cited examples of such digital-labor-derived value center on Facebook and the Huffington Post (HuffPost). For Facebook, the inherent value of the website rests solely on the content created by all its users, which is instantly posted without compensation. For the Huffington Post, its journalistic content was created by volunteer-journalists, unpaid individuals whose editorials eventually enabled the sale of Huffington Post to AOL for $315 million. Upon the sale of HuffPost, however, zero dollars went to the journalists and writers because they were volunteers. They were members of a new digital-laboring class.

A third and final example of digital labor worth studying is that of reCAPTCHA. Originally designed, ironically, to detect whether a purchaser of online tickets is in fact a human rather than a computer program hoarding tickets for resale, the developers of CAPTCHA attempted to find activities that would distinguish human behavior from AI behavior. Because AI is a moving target, CAPTCHA needed to constantly shift its humanity-detection goalposts as AI became ever more capable at mimicking human skills. One particularly effective, historical version of CAPTCHA was to present a warped word or number sequence on the computer screen, asking the human to type in the correct word or number.

This triggered a professor at Carnegie Mellon University, Luis von Ahn, to consider monetizing the massive "character recognition" effort demonstrated by millions of purchaser-citizens daily. ReCAPTCHA invented a new, digital-laborer-powered three-way deal: The *New York Times* wished to digitize its entire paper archive. But many old newspapers had smudges and blurs that, when run through optical-character-recognition systems, the computer failed to read. Ticketing companies wished to disambiguate human purchasers from computational robots. ReCAPTCHA made a deal with each. In lieu of presenting digitally constructed, blurred-out word phrases to purchasers, reCAPTCHA presents precisely the newspaper line images that the computers failed to digitize and asks purchasers to type in the correct words. If a purchaser agrees with other purchasers, then she is human. And reCAPTCHA would make money doubly, by selling manually digitized words back to the *New York Times* and by selling a human-detection system to Ticketmaster. As for the ticket purchasers, they are not compensated for their efforts and time; rather, they derive the *pleasure* of having a chance to spend money on a ticket to a show.

Two examples of reCAPTCHA-based digital-laboring strategies are shown in the frontispiece to this chapter. After several years of effort, reCAPTCHA completed the digital correction of archival *New York Times* print editions, and these two examples represent other creative ways in which the human populace has been enlisted into providing three-way value for corporations. Figure 4.1a shows a human-verification system that uses not warped words from a digital scanner, but the act of reading numbers on photographs of homes. These are real-world photos of homes on streets across America, and the value generated by this identification act is to help digital map systems identify the exact location of residential home addresses on national maps.

Figure 4.1b shows an example of *image classification*, in which the human user must identify every picture containing a specific type of object—in this case, street signs. This example is both rich and ironic, because the human purchaser is doing work to create massive labeled examples of categories of objects. Once hundreds of people have clearly identified pictures that do and do not have signs in them, then corporations owning that labeled information and owning the images can use the two sets to create deep learning neural networks that can successfully find signs in billions of images. With these new algorithms in hand, a corporation would have the power to find all the signs in the images taken by roving photographic automobiles throughout the world, digitizing signs—essential to the self-driving-car revolution. Thus digital labor performed by humans ends up aiding AI's own advancement in this case, leading eventually, through a rather complex cascade, to monetization for the owners of images and AI solutions down the road, literally displacing yet another human skill.

Digital labor inverts the value relationship between corporation and consumer by making the consumer into the producer of digital goods. The original producer becomes the owner of capital—the AI capital that makes a new machinery of labor, profit, and monetization run forward in a world architecture that is ruled not by the exchange of goods but by the exchange of information. Information, the ownership of which is distinctly material to the enslavement of humans, plays a new role today in the exploitation of many for the incremental benefits afforded to capital owners, completing, full-circle, the ways in which new and old technology alike bear upon manifestations of labor and labor exploitation in our human society.

In his recent book, *Surveillance Valley*, investigative journalist Yasha Levine uncovers the lines of funding and influence between military and intelligence government funding and the birthing of the internet. His remarkable research demonstrates direct surveillance-oriented collaboration, not only between major internet corporations and government surveillance, but also in how the orientation of technology innovation toward comprehensive human behavior tracking is a direct evolutionary descendant of military intelligence and counterinsurgency efforts: "For many Internet companies, including Google and Facebook, surveillance is *the* business model. It is the base on which their corporate and economic power rests. Disentangle surveillance and profit, and these companies would collapse. Limit data collection, and the companies would see investors flee and their stock prices plummet."[26]

The generation and ownership of digital information, more and more pervasive throughout society, is a clear line of influence on the *unequal* in society. At the same time, the very same digital information is aiding AI in becoming ever more capable at besting human capability. As AI progresses, due in no small part to human digital labor, it challenges our understanding of *human* by both diminishing our uniqueness in privileging the *nonhuman* and raising the specter of a more capable evolutionary progeny that is *transhuman*.

Discussion Questions

1. What is emotional labor? How does inequality specifically introduce the business case for emotional labor? How does advanced technology, such as surveillance in the workplace, exacerbate the deployment of emotional labor?

2. Does a critical understanding of inequality in society extend to AI? To be specific, does the hypothesis that AI may achieve a level of intelligence *unequal* to that of humans have impact on societal inequality? How?

3. Stephen Hawking said, "The rise of powerful AI will either be the best or the worst thing ever to happen to humanity." What are specific examples of the best and worst cases?

4. How can digital labor increase equity and quality of life? What regulatory or ethical mechanisms would be needed to create appropriate guide rails?

5 Surveillance, Information, Network

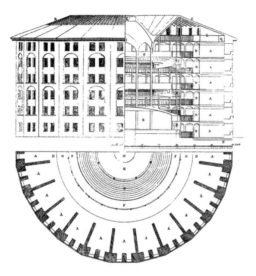

Figure 5.1
Bentham's panopticon, illustrated by Willey Reveley (1791).

He's Dead!

It's software. It mimics him. You give it someone's name. It goes back and reads through all the things they've ever said online; their Facebook updates, their Tweets. Anything public. I just gave it Ash's name, the system did the rest.

It's Sick!

Just say hello to it. If you like it, you then give it access to his private emails. The more it has, the more it's him.[1]

—*Black Mirror*, "Be Right Back"

Sources

Keywords for Today: Information, Network
Ciaran Carson's "Intelligence," in *Belfast Confetti*
Minority Report
Black Mirror, "Be Right Back"

Guiding Issues

Humans are intimately connected to information creation. As you saw in chapter 4, human action itself is now monetized as *digital labor*, connecting daily activities of communicating, purchasing, and exploring with information that translates directly into corporate profit on a scale previously unmatched. This chapter encourages you to explore the source and destination of digital information, from surveillance to data acquisition to information processing to global information networks. That information *lifecycle* will help us to understand how and why modern information influences networks of power in society and thereby challenges or reasserts features of our humanity and culture.

Artificial intelligence presents a new variable in the information and society interplay. AI is a consumer of information, and with information it can produce new forms and varieties of knowledge that were previously unreachable. An interactive digital advertisement presents you with an ad and a discount code for a new coat. But first it estimates your societal class, gender, and age using computer vision. Next, it measures your flush response, heart rate, and pupil dilation as the advertisement is presented. The AI system behind this interactive billboard is not simply a passive information collector; it builds a model for advertising efficacy, experimenting with millions of variations across thousands of human archetypes. Your reactions are yet more digital labor, providing a new, experimentally validated sales model to the corporation that buys the collected, tested, tweaked, and optimized purchasing-behavior data. But how might we classify all the information that is gathered, formulated, and then repackaged to output a tailored advertisement? How is that information gathered and shared by you, the consumer; by the advertiser; by the company that assembles the coat; and so on?

This chapter asks: How does AI, coupled with the new information lifecycles that vary from surveillance systems to networks, impact or influence elements of our humanity, our culture, and our relative power systems? As we explore these questions pertaining to information, historical and fictional representations of surveillance as a means for gathering information (and, in turn, its impact on humanity) will be critical

to our lines of inquiry. We will attend to examples ranging from Bentham's panopticon to Ciaran Carson's "Intelligence" to *Black Mirror* episodes like "Be Right Back." As
we attend to these examples, we will consider what information looks like and how
it circulates and determine how the varied forms and circulation of information can
influence our relationships with one another, our institutions, and the societies that we
build and with which we engage. In *Keywords for Today*, MacCabe and Yanacek indicate
the multifaceted nature of information in our current times and how this has evolved
in significant ways just in the last century:

> Among other uses, information refers to rapid social transformation associated with digital
> technologies and global communications (especially in late twentieth century compounds
> such as information superhighway, information age, and information revolution). Yet because
> of information's range of technical and general senses, in many circumstances it remains
> unclear exactly what information is. What information means, accordingly, is both a prac-
> tical problem (for example in assessing who the rightful owners, controllers or appropriate
> recipients are for whatever is called information) and also a wider social question: what kind
> of economic, regulatory, or cultural information regime will allow societies to secure appropri-
> ate access to information as an essential quality of civil society and democracy, regardless of
> whether they are, in recent "digital divide" thinking, information-rich or information-poor?[2]

Two modern films provide particularly lucid futuring exercises for the relationship
between information and humanity. *Minority Report* enables us to attend to the question of information and determinism: How does information increase our modeled
understanding of the future, and how does such deterministic prediction in turn influence our sense of personal agency and free will? With the *Black Mirror* episode "Be Right
Back," we turn our attention to data and identity: Does the accumulation of data itself
approximate reality and human identity? Is a person more than the accumulation of
their vast digital footprint over time? This episode elegantly explores all sides of this
constellation, turning back to the question: What is most essential in being human?

We turn to IBM's question-answering AI system Watson as an example of an evolving
network. In this example of an AI laborer, we see its expertise through access to vast networks of specialized information, often achieving super-human levels of efficacy in fields
as varied as the *Jeopardy!* game show, medical diagnoses, and financial recommendations.
We deconstruct the Watson question-answering system, demonstrating exactly how
human computational architecture, AI statistical power, and networked data combine to
effect sophisticated levels of machine performance. This particular example allows us to
interrogate how the linking of information, based on its categorical features, allows us to
develop new categories for understanding information and how its links to other forms
of information may uniquely define our particular moment in human history.

Along the way, we can study how the essence of such an AI-driven expert system is fundamentally alien to how human decision-making occurs. We also consider how information is gathered, configured, and often mobilized as a means for shaping, controlling, or disenfranchising individuals or populations. Such illustrations are offered in Ciaran Carson's poem "Intelligence" as examples of how we exercise control over the dissemination of information or fall subject to powers beyond our control that collect, forcibly extract, or broadcast information that individuals or groups might prefer to protect or to keep private.[3]

Language: Information and Network

Recent feverish excitement about deep learning emphasizes its connection with data: it is the billions and trillions of exemplar datasets available online that have enabled AI neural-net-based computational systems to perform at levels that were unimaginable just one decade earlier. Indeed, data is the driver of numerous major, recent AI advances. And these vast datasets are one iteration of how we package or store information in our current moment. Relationships between AI and data are more nuanced, perhaps, than our historic relationships between humans and information. AI consumes data, but also produces new information. That information, in turn, has new properties because it is digital, networked, and pervasive. In this chapter, we will interrogate AI's relationship to humanity by considering the role and lifecycle of information and how the development of networks in particular reshapes and redefines our historic relationships to information.

To begin this inquiry, we recognize that information has a source. Surveillance generates data, and it is just one of many approaches to data collection. It will be critical to this chapter, however, because surveillance touches upon agency, power, and scaling as a data-collection methodology that can grow explosively thanks to the lubricating effects of computational technologies. Computers, in turn, are the only systems capable of taking the massive data gathered thereby and converting it into semantically meaningful *information*, which betrays the particular variation of information as it exists in our current world (in addition to how it is gathered). What information means to society and how it is stored are central issues in this examination—particularly because the words themselves, *information* and *network*, have both a rich history and a pertinent, recent evolution in cultural meaning and influence.

The historical origins of *information* relate to the verb *inform*, often relating directly to the act of teaching, instructing, and educating. But in modern senses, the focus shifts from the process of informing to the product that results from that act. As a currency of knowledge, information becomes something one can have more or less of, and as such it is a societal discriminant irrespective of how it was developed, gathered, computed, or derived. In *Keywords for Today*, we see the fissures in the taxonomy of information that are evident in the evolution of the word and are indicated in the "wider social questions" associated with "what kind of economic, regulatory, or cultural information regime will allow societies to secure appropriate access to information as an essential quality of civil society and democracy, regardless of whether they are, in recent 'digital divide' thinking, information-rich or information-poor."[4]

The very existence of computational systems that produce and consume information also divorce information (in its various forms) from being an exclusive province of humans and human culture: "This emergent information technology meaning presents information as something abstracted from any human practice of 'informing' or 'being informed.' Instead, information is a characteristic of any arrangement of signs that can be stored, viewed as a message, and transmitted in signal form; so it can—and increasingly does—pass between inanimate devices as much as between animate beings, as well as through combinations of the two."[5]

The digital divide extends beyond the classification of humans because information is a currency that is transhuman. In the new world of digital information, humans, corporations, and AI systems can produce, digest, own, monetize, and transmit information. Thus information itself can be seen in modern times as a concept that imposes class distinction, and does so across human and nonhuman agents throughout society. This evolution of information, thanks to digital technologies, acts both as a grand equalizer between agents, blurring away distinctions in the affordances of humans, corporations, governments, and machines, and simultaneously as a creator of new inequalities, cutting into human and non-human-agent categories agnostically. The various forms of information and the creative and sometimes still unimagined ways in which information can be configured to have empowering and disempowering effects on human individuals and populations raises anxieties in individuals and collectives of humans alike.

As machines transform data into value-laden information, that information must be stored, shared, and communicated to actualize its inherent value. That architecture for ownership and sharing is termed a *network*. These networks can signify a hierarchy or nonhierarchical architecture of connectivity. In *Keywords for Today*, MacCabe and Yanacek suggest that *network* is a keyword particularly of our age. They write:

> The vast majority of keywords in English, and of its conceptual vocabulary more generally, derive from Latin. Network, however derives from two words central to the vocabulary of Old English and both of which are used frequently in contemporary language: *net*, a piece of openwork fabric forming meshes of a suitable size to catch fish and *work*, whose primary meaning

is labor or activity. While only of specialized use in Early Modern and Modern English, in the last fifty years network has become central to the vocabulary of transport, telecommunications and computers and has developed meanings so crucial that there are serious claims that we are living in a network society.[6]

As we consider the influence of technological advancement on humanity, few words in English better capture the current state of affairs. If we "are living in a network society," then the centrality of links between our language and the cultural norms, practices, and economic underpinnings that feature in our everyday lives are present in our various networks that we both manage and inhabit. There are various networks that act upon us as well, however.

The "openwork fabric forming meshes of suitable size to catch fish" and the primary meaning of "labor or activity" lay a baseline for the societal linking of utility and interpersonal relationships that are part of the various "networks" that are foundational to our networked society. If we are to parse through the taxonomy of literal and figurative networks in a society increasingly influenced by artificial intelligent systems, we have a rich cross-section of materials to consider as we analyze the potential benefits and costs of this new human age. As MacCabe and Yanacek note in citing Bourgouin, the structures inherent to culture, community, and class lose their discriminating potential in our new, network-centered model of the world: "'What used to be easily referred to as relations within a "culture," "community," "class," or "group," is now indiscriminately being called a network' (Bourgouin, 2009)."[7]

The modern network has power through its information architecture, and in so flexing its power, it has the potential to diminish the relative power of other forms of social relationship and social communication. AI operates not in the human world of culture, but in this ethereal realm of the network. As such, it has agility and responsiveness within a worldview in which we humans have no particular advantage or experience.

The Panopticon

Just as technology writ large has had millennial impact on war-making, its development has been particularly poignant in the mobilization of *surveillance*—the ability of a group in power to establish close watch over others. But the application of technology presaging artificial intelligence to surveillance is particularly enlightening because, surprisingly, AI's roots are in fact borne out of the rudimentary surveillance needs of World War II. Alan Turing considered the problem of breaking the German Enigma code to assist the Allies. His breakthrough was the invention of a computing machine that could consider possible combinations of code-breaking solutions at a speed greatly exceeding that of human code-breakers. The labor of spying was reduced to machinery through Turing's invention and arguably shortened the war by several years.

The computations undergirding Turing's physical machine became the nascent science of computing. In formulating a way for machines to "innovate," to solve problems with unknown solutions, he laid a foundational vision for the modern, intelligent computer. But a parallel effort in surveillance technology, borne of an even earlier time, grew to meet up, postwar, with Turing's promise of a general artificial intelligence. Its origins stretch back further, to the eighteenth century, when Jeremy Bentham proposed a new architectural design for English prisons. Prisons have derived their lasting control over inmate populations chiefly through the often brutal assertion of hegemonic power. It is in the threats and displays of punishment meted out to prisoners that a social order becomes forged and reforged as a constant reminder of the position of the disempowerment of each inmate.

Jeremy Bentham proposed a more nuanced form of power hegemony—one that is based explicitly on surveillance and, significantly, on prisoners' explicit awareness of that surveillance.[8] His architectural proposal for a prison (and, indeed, for schools, hospitals, and many other institutional buildings) consisted of concentric cells placed around a central watchtower subject to a geometric constraint: that a single guard in the watchtower could observe any inmate's cell at any time from that one central location. Bentham recognized the power of surveillance, not to obtain explicit information, as in Turing's World War II efforts, but to establish a generalized behavioral modification of the imprisoned population with surveillance made a constant, never-ending threat. Surveillance becomes the less violent (yet almost equally efficacious) cousin of brutal force in a prison system. Bentham called the panopticon "a new mode of obtaining power of mind over mind, in a quantity hitherto without example." Echoes of this form of public surveillance are evident even in the newest AI-based military technologies, as in drones that "loiter." They provide constant reminders to a civilian population that it is under watch by powers beyond its reckoning.[9]

Earlier forms of such military tools are evident in former conflict zones like Northern Ireland. In Ciaran Carson's "Intelligence," we are offered poetic insights into the psychological implications of a watchful eye on a subject who knows little of what personal or community-based information is being gathered, extracted, configured, and networked for parties that may or may not have ill intent. In the context of Carson's poem, the narrator is caught in the midst of a national crisis in a historically disputed area. Both the Irish government and the British government lay political and geographical claims to a region the population of which has aligned with one national government or the other over the course of centuries. In the fever-pitch era of conflict from

1968 to 1998, Carson's poem captures the cutting-edge technological tools that menace a civilian population caught up in military and paramilitary conflict. His narrator captures the unease associated with sophisticated and rudimentary tools used to facilitate the gathering and configuration of information from a civilian population caught in the crosshairs of the military and political agendas that swept the region. He writes:

> We are all being watched through peep-holes, one-
> way mirrors, security cameras, talked about on walkie-
> talkies, car 'phones, Pye Pocketfones; and as this helicop-
> ter chainsaws overhead, I pull back the curtains down
> here in the terraces to watch its pencil-beam of light flick
> through the card-index—*I see the moon and the moon sees me*,
> this 30,000,000 candlepower gimbal-mounted Nitesun by
> which the operator can observe undetected, with his infra-
> red goggles and an IR filter on the light-source. Everyone
> is watching someone, everyone wants to know what's
> coming next, so the lightweight, transparent shield was a
> vast improvement over the earlier metal one because
> visibility was greatly increased—an extra bonus—
> gave better protection against petrol and acid bombs
> which could flow through the grill mesh of the metal
> type.[10]

As he delineates the variety of tools used by those in power (whether military, government, or paramilitary) that range from peep-holes and one-way mirrors to costly and sophisticated infrared goggles or a later mentioned "Telescope Starlight II LIEI 'Twiggy' Night Observation Device," Carson's narrator demonstrates how negotiations of power are at play beyond the simple breakdown of who gathers information and who uses information against another. The crystalized distinctions in these power positions are exemplified in access to crude and sophisticated tools to uphold, solidify, and calcify these positions in a power dynamic through surveillance or, as he puts it, the information gathered in "Intelligence."

The narrator in the poem articulates vulnerability in these uneven positions in his relationship to the police and military who operate the helicopter with its "Nitesun" searchlight beam that wields thirty million candlepower over the night sky, illuminating the terraces below. Carson writes, "I pull back the curtains down here in the terraces to watch its pencil-beam of light flick through the card-index—*I see the moon and the moon sees me*." Invoking a position, "down here," and opening drapes in the confines of a home with the menacing sound of the "chainsaw" helicopter motor overhead, the narrator echoes the vulnerability of childhood with the nursery rhyme refrain, "I see the moon and the moon sees me." A configuration of sophisticated tools, simple location (high vs. low), and language that articulates infantilizing sentiments, vulnerability, and a threat outside, Carson's protagonist captures the psyche of an oppressed population, the information of which is extracted, configured, and shared while its members have little power to control or shape the "intelligence" that is gathered.

Published in 1989, "Intelligence" is part of the *Belfast Confetti* collection. This poem in particular centers on the sense of vulnerability and pervasive threat that an individual or community can feel in a societal configuration modeled after Bentham's panopticon. Carson writes, "Keeping people out and keeping people in, we are prisoners or officers in Bentham's *Panopticon*, except sorting out who's who is a problem for the naïve user, and some compilers are inclined to choke on the mixed mode—panopticons within panopticons."[11]

In Carson's Belfast, unsophisticated surveillance systems are afforded through individual citizens—informers—who are willing to spy on and report to various powerbrokers in a deeply divided and fissured society. Informers could work for undercover officers for the Royal Ulster Constabulary or the Irish Republic Gardai, operators for MI6, or the paramilitary organizations that spy on the civilian population. Anyone could be a prisoner or an officer in the panopticon at any given time, roles switching and shifting, depending on the community context. In an emergency state, it is difficult to discern who is watched and who is watching in "the mixed mode—panopticons within panopticons."

Yet certain elements of the powerful and the powerless were obvious too. In the opening stanzas that delineate tools for surveillance, from the seemingly innocuous peephole and car phone to the "2B298 surveillance radar [that] can identify moving man at 5,000 meters by the blips on its console," advancing technology that is weaponized against a civilian population by its government and military is terrifying at best. But that is also the point. The tools offer another arms race to control a population, to yield the compliance that a government or policing service in an emergency state seeks. Carson writes, "Everyone is watching someone, everyone wants to know what's coming next, so the lightweight, transparent shield was a vast improvement over the earlier metal one because visibility was greatly increased—an extra bonus—gave better protection against petrol and acid bombs which could flow through the grill mesh of the metal type."[12]

As he describes the fissures in a society rocked by nationalist politics and violence in the street, he sets up the parallels in perspective in the context of a Saracen, the state's police vehicles. The civilian and the paramilitary operator cannot tell by design where the Saracen are watching for the next petrol or acid bomb "which could flow through the grill mesh" of the metal shield. But if the operator can indeed see through a "transparent shield" rather than the "grill mesh of the metal type," who will gain the advantage? With the advancing technology of the transparent shield to at once protect but also clarify the viewing field for the Saracen operator, which of the individuals or communities in strife will have the advantage to more quickly learn what's coming next?

Wars on Terror and Modern Surveillance

Fast-forward to September 11, 2001, when attacks on the World Trade Center and the Pentagon in the United States triggered a new vigor for the use of surveillance and its careful configuration and networking. In the context of military, governmental, and local policing in nationalist political unrest predating 2001, operators could be divided into civilian, military, paramilitary, policing, and governmental categories, among others. How individuals or communities shifted from one category to the next or often inhabited multiple categories was highly complex and subjective. That said, the non-state leveraging of violence for political means in the context of Al Qaeda and its more recent heir, ISIS, has driven the development of more sophisticated tools for gathering information by governmental and military agencies. Even more complex and sophisticated formats for configuring such information into various networks to be leveraged in a myriad of ways by various power players in this new world political dynamic have been made possible through advancing technologies that build on and assist in the sophistication and configuration of these networks. The more sophisticated tools have also led to more sophisticated legislation to protect their use by governments the world over.

The US Congressional bill designed to authorize previously unimaginable levels of domestic surveillance, the Homeland Security Act, was debated very quickly because of the mood of urgency pervading Capitol Hill. But pundits in the press began to notice just how thoroughly the panopticon concept was about to go digital in 2002. William Safire wrote an opinion piece in the *New York Times* in November 2002, with words that were to become prescient about just how thoroughly digital surveillance technologies could bring Orwell's telescreen[13] to life:

> Every purchase you make with a credit card, every magazine subscription you buy and medical prescription you fill, every Web site you visit and e-mail you send or receive, every academic grade you receive, every bank deposit you make, every trip you book and every event you attend—all these transactions and communications will go into what the Defense Department describes as a virtual, centralized grand database. To this computerized dossier on your private life from commercial sources, add every piece of information that government has about you—passport application, driver's license and toll records, judicial and divorce records, complaints from nosy neighbors to the F.B.I., your lifetime paper trail plus the latest hidden camera surveillance—and you have the super snoop's dream: a Total Information Awareness about every U.S. citizen.[14]

The Homeland Security Act passed nonetheless, and an age of dark ops, computational surveillance over all US citizens, began, only to be publicized in 2013 due to revelations made by Edward Snowden of the National Security Agency (NSA). Today, we know from Congressional testimony that all phone records of all US citizens are stored and that semantic-analysis programs evaluate every email message, considering and categorizing messages as potentially harmful based on word and sentence content. The computational technologies of Turing have matured into AI systems capable of performing surveillance on everyone simultaneously. Bentham's vision of visibility through architecture has become visibility through the internet architecture—and, as Bentham presaged, at a scale that was previously inconceivable. London has a network of more than 420,000 CCTV cameras, affording continuous monitoring of nearly every square yard of the city.[15] Machine algorithms evaluate these images in the laboratory, with the hopes that one day soon AI systems will be able to view all 420,000 video streams in real time and trigger real-time responses in turn.

Turning to the present state of play in technology innovation and surveillance, computer science has taken up positions on both sides of an escalating battle: advanced encryption techniques to defeat surveillance, facing off against advanced analytic techniques to preserve surveillance. Strong encryption continues to enjoy wider use, even in existing applications, such as Gmail. But research into quantum computing has the potential to render secure email and secure web page encryption meaningless, assuring the wealthiest actors in society (governments and corporations) the power to collect the most encrypted of information sources. Whether we live in a democratic country or a nation ruled by a dictator, continuous surveillance of our lives has become a fact— whether by businesses for the sake of monetization or by governments to guarantee power dominance. Technology, and in particular artificial intelligence, has become central to the power dynamics of surveillance. More than any other case we can identify, this is where science fiction and reality have come to merge into a nearly indistinguishable union in the present state of the art.

After Data

Surveillance is both a psychological tool and a process for creating value. We now turn to that value-creation process, which plants the fruits of surveillance, data, in a garden that produces new knowledge: the network. Although AI has the power to greatly improve the quantity and quality of data captured worldwide, its most significant edge is displayed in how that data is stored, mined, and transformed into value and power. We use two films as futuring exercises that interrogate the boundaries of critical ways in which that data, thanks to advanced artificial intelligence, might change culture. *Minority Report* considers the possibility of pervasive data collection, and through the trope of extrasensory perception, it further imagines a world in which the future is deterministic.[16] Never far from works of science fiction, including Isaac Asimov's classic concept of *psychohistory* presented in the *Foundation* trilogy, the basic idea behind data-informed determinism is that with sufficient information and computational resources, perhaps we can truly predict the future, by reducing every action, every decision, to the parameters, variables, and values that govern the underlying circuitry of human lives. *Minority Report* is itself based on Philip K. Dick's 1956 short story, "The Minority Report," which also directly explored questions of agency, free will, and determinism. We will later engage in the controversy of determinism itself—whether rational engineering can explain the behavior of a human being as a derivation of our smallest parts. But without attending to that debate, what *Minority Report* offers is a vision of societal power relationships when predictive power is concentrated in the hands of the owners of information.

In a fully networked system, nearly all data becomes actionable intelligence. At one extreme, stores can provide bespoke advertising based on prior purchasing history, creating apparent value for the consumer and producer both. At the other extreme, a police force can locate anyone, anytime, and can interrupt crime before it is committed, creating optimized societal safety. The individual reduces from an actor of free will to another data point—one to be not only measured, but manipulated, like an electron in a quantum physics experiment. In *Minority Report*, the dystopian depiction suggests that safety is afforded through governmental monitoring of all individual citizens' data. Much like the underpinnings that rationalized the Homeland Security legislative changes in 2002, wide-scale gathering of data through surveillance of a nation's population is rationalized and justified in the name of predictive policing. The notion of preventing crime through the policing of intention, not just action, touches on the societal buy-in that needs to be combined with fictitious tools that make such coordination of data, policing, and political will a possible foundational feature of a society. Although the dystopian film suggests all that can go awry with such a narrative, we also know in our present context that the tools are now advanced enough and the data sources rich enough that this fantastical film is not quite so far-fetched as it was in post-9/11 America when it debuted in 2002.

In the *Black Mirror* episode "Be Right Back," we observe a more deeply personal identity challenge in the face of near-infinite accumulations of personal data in public networks: What are the features of difference between a person and their digital trail? How is privacy navigated when individuals volunteer their information in social network systems? What information is private, what is public, and what can be bought, sold, or usurped in the name of safety or to alleviate suffering, as is depicted in the case of grieving as suggested in "Be Right Back"? These questions are not purely the prov-ince of science fiction any longer when we consider the vast data sources that are both volunteered and coerced presently in social networking, digital records of consumerism and vast holdings of mobile phone records held by agencies like the NSA. Directly rel-evant to Charlie Booker's existential questions in "Be Right Back," Raymond Kurzweil, one of the major proponents of human immortality, claims that he is *already* effectively immortal. When asked how this can be true in a conference panel, Kurzweil responded that his digital footprint is so comprehensive already that a future AI will be able to recreate him. This is not fiction writing; it is the semantic argument of a world-famous scientist devoted to life extension.[17]

In "Be Right Back," Charlie Brooker investigates the boundaries of digital simulation: Is emotion, expressed in tones and timbres identical to a person's true style, the same affect, or is it artificial? Is companionship with a data-networked, intelligent reconstruc-tion true companionship with fractional value, or is it worth little more than other relics of a life, like family photo albums? The *Black Mirror* episode leaves us in a boundary space at the end, with Martha unwilling to extinguish the simulated life of her simulated boy-friend, Ash, and equally unwilling to grant it human rights and responsibilities.

Brooker suggests that the networks of information that we leave in digital form con-tend with the complexity of lived memories in those with whom we build relationships and lives. The episode suggests that humanity is more than information, data, and the configuration thereof. AI provides us with a lens through which to question the value of life as uncoupled from the value of actions taken, scenes witnessed, and emotions felt. Its machinations at simulating each tendril of human character will only grow over time, and with this progress in time, our understanding of human identity and human experience will continue to be challenged by simulacra driven by vast networks of data and experience.

Watson: A Case Study in Deep Data

The worlds supposed by "Be Right Back" and *Minority Report* reside in imagined, far-flung futures. If we turn our attention next to the power of *present* AI-based data-understanding systems, a deep dive into just how they operate can elucidate the ways in which these systems, already capable of significantly reconsidering categories of human work, oper-ate along methodologies that are remarkably alien to the manner in which humans have historically absorbed data and used networks for decision-making purposes. Wat-son, the family of AI-based question-answering systems marketed by IBM, represents the state of the art in this field of AI-powered machines that exhibit high levels of human skill. Watson came to prominence in 2011, when a *Jeopardy!*-playing version of the computer architecture beat Brad Rutter and Ken Jennings, the legendary human *Jeopardy!* champions, in a series of highly publicized television matches, winning the grand prize of one million dollars.

In the search for commercial monetization, IBM followed Watson's *Jeopardy!* win by investing significant resources in creating specialized versions of the architecture, starting with one suitable for medical diagnosis across a broad range of medical specialties and conditions. In applying AI question-answering technology to the medical domain, IBM directly challenged preconceptions regarding the human role in a field known to be highly complex and requiring peak human educational attainment. Physicians themselves, faced with the success of Watson at diagnosing human disease from symptoms with very high accuracy as compared to their own peers, have gone on the record to herald the machine's application in medicine and to recognize the dawn of a new medical age. These are physicians, not computer scientists, who have watched Dr. Watson's behavior and are convinced that AI's role in medicine is genuinely revolutionary. Satoru Miyano, a professor of the Human Genome Center at the Institute of Medical Science in the University of Tokyo, commented to Sharon Gaudin in *Computerworld* about the relative power of computer-based diagnosis: "Miyano said researchers and doctors are faced with too much data. Last year, he said, more than 200,000 papers were published about cancer alone. Meanwhile, 4 million cancer mutations also were reported. 'Nobody can read it all,' Miyano said. 'We feel we are a frog in the bottom of the well. Understanding cancer is beyond a human being's ability, but Watson can read, understand and learn. Why not use it?'"[18]

As reported by Jon Gertner for *Fast Company*, Watson is perceived to have so high a diagnostic accuracy that the power relationship between the system and its human minders, the former top-flight diagnosticians, is changing dramatically:

> Watson can ingest more data in a day than any human could in a lifetime. It can read all of the world's medical journals in less time than it takes a physician to drink a cup of coffee. All at once, it can peruse patient histories; keep an eye on the latest drug trials; stay apprised of the potency of new therapies; and hew closely to state-of-the-art guidelines that help doctors choose the best treatments. Watson never goes on vacation. And it never forgets a fact. On the contrary, it keeps learning.
>
> Kris shows me what happens when Watson's treatment plan calls for an MRI. A button pops up on his screen to ask for preauthorization. "I just click that," he says, and it's done instantly.
>
> I ask him what if Watson's request is denied.
>
> Kris seems amused by the question. Watson has already consulted the latest medical literature, and it's been trained by the best cancer doctors in the world. "Who is the authority that is going to trump that?" he asks. Insurers balk at paying for unnecessary procedures; Watson's expert opinion essentially guarantees the necessity.[19]

To the medical practitioner naively reading early reports of field successes, Watson bested humans by nearly every standard. This form of technological acceptance will likely occur in disparate categories of work in which AI systems will use data and networks to make recommendations and, ultimately, decisions. But it is equally important to appreciate just how these AI systems actually operate, as contrasted to the manner in which human practitioners use data, experience, and problem-solving techniques to make decisions using human cognition. Early field success, after all, can be a mixture of optimism and hyperbole, as has been made very clear through recent investigative journalism devoted to understanding just why Watson has not become a common diagnostician throughout medicine.[20] We must dive into the details of Watson's architecture.

Deep Question-Answering: An Architecture in Detail

The design goal of a deep question-answering (DeepQA) algorithm is to be able to use a massive corpus of information, ideally online and ever-changing, to select the likeliest answer to each posed question and to also identify its own level of confidence in the chosen answer. Two steps are required to arrive at a working DeepQA system: (1) implementing a specific system architecture and then (2) running statistical processes to convert a specific question into a confidence-labeled answer.

Architecting the System

The architecting process demands the skills of technical computer scientists working in concert with domain experts who understand the existing body of published literature that the system must tap. Seven core activities comprise the architecting process, each dedicated to designing a connection between stored data and expected question types so that real-time search can yield the most relevant possible evidence supporting the answers that will be considered:

1. *Domain specification.* Decide on the narrowest possible definition of the range of topics in which the DeepQA system is intended to operate. The narrower the specified domain, the smaller the overall corpus of data required for the system to achieve highly competent question-answering and the likelier that the system may be able to approach or even outstrip human capability. Note the tension in this step in contrast to how liberal arts human education proceeds, by hypothesizing that broad educational exposure to many disciplinary content areas empowers a person to think laterally, thereby incorporating creativity and innovation into any narrow discipline. Here the measure of power for humans and for computers splits along the axis of breadth: interdisciplinary innovation for humans; highly focused, deep data for computer expertise.

2. *Question analysis.* Characterize the set of possible question styles that will be presented to the system. This is often done manually by having researchers and content experts collaborate to collect massive numbers of example questions that match the sort of questions we wish the DeepQA system to answer. In the case of *Jeopardy!*, this task was accomplished readily by collecting every historical question ever asked in the game show. In the case of prostate cancer, medical consultants must construct templates, or taxonomies, of all the specific forms of questions that will be posed by diagnosticians to Watson. This step, once again, is in fact a narrowing exercise; the architecture ultimately will be optimized for the chosen question types, and *only* for those questions. This reduces the complexity of the field over which the system must perform, superspecializing the DeepQA as much as possible to maximize its performance in a narrow space while discarding concerns about its performance outside that specified domain of question space. The human ability to encounter and consider previously unconsidered problems or to operate elegantly when the problem evolves from an expected to an unexpected form is explicitly thrown out of consideration. The concepts of *robustness* and *resilience* are often used to characterize the performance of a person or machine in the face of failure, uncertainty, or unexpected situational change. These properties, so important for an airplane pilot and an emergency room physician, are being explicitly eliminated from the architecture to maximize machine performance under nominal, ideal conditions.

3. *Content acquisition.* The next step is to gather all the digital content that will be used as reference material in which the answers to all narrowed questions will reside. Content expert researchers manually choose all possible sources initially. For the *Jeopardy!* version of Watson, such sources included encyclopedias, thesauri, dictionaries, internet sites, literary works, and so on. A superhuman effort can be involved in formulating a list of all such information sources, then identifying digitally accessible versions of the information, then retrieving and storing rapidly searchable versions for the following architecting steps.

4. *Automatic content expansion.* The acquired content is not the end-all information corpus; rather, it is a starting point in the development of an information network search that will enable the machinery of online computing engines to broaden the information corpus by orders of magnitude. The computer begins by running statistical analysis on the sources that humans have chosen—for instance, looking at words frequently appearing in the dictionaries and literary works chosen. Using these words or phrases as indices, the computer performs a general internet-based search, identifying high hit-rate matches in any and all online works. If the machine finds content online that matches highly with existing sources, then the online

content is added to the body of sources, expanding the overall content available and triggering this entire indexing and online searching process to run again. This data-expansion process, iterative and broad-ranging, creates the massive data corpus that is critical to the potential success of DeepQA systems. Indeed, any deep learning or deep analysis system requires extreme amounts of data, and this semiautomated step is the key to producing that scale of information. Here we see a significant departure in how machines become set up for question-answering as compared to human experts. Humans do literature surveys, to be sure, but we have no capacity for billions of documents. Instead we rely on nuanced relevance-reasoning, abstraction, and analogical-reasoning methods that help us formulate hypotheses that find patterns in sparser data in broad-ranging ways. The computer, easily viewed in this circumstance as a statistical calculator, does not conduct such elevation from sparse example to general theory. Quite the contrary: the computer system desires every possible piece of complementary data possible. It has no imagination, no abstraction, no induction; rather, it has precision, recall, and nearly unlimited memory capacity.

5. *Question-analysis tools.* Next, engineers design dozens or even hundreds of small algorithms that transform each question into a list of key aspects relating to the question. These diverse aspects can include a list of key concepts most relevant to the question; categorization of the question type; and many other formulations of the question's mode, type, and nature. Any technique that will sometimes prove beneficial is considered good enough to be worth adding because the underlying goal is to break a question down into as many different formulations of its content and type as possible.

6. *Hypothesis search tools.* The question-analysis tools generate keywords, tags, question types, and many other categorical labels for each and every question. Next, engineers decide how each such keywords, tags, or question types will be used to conduct a broad search across the expanded content (content expansion's results) in a massive fishing expedition for possible answers. Techniques frequently used at this point include word search, synonym search, phrase search, and statistical word phrase techniques. Many search strategies are created, not just one, with the goal of transforming a question's categorization into the largest, most diverse possible set of "hits" across expanded content as possible. This is the digital analogue of the concept of *casting a wide net*—and in the case of DeepQA, that wide net can yield literally millions or billions of possible elements of evidence that may prove relevant to the question at hand.

7. *Evidence retrieval scorer.* Finally, engineers must also architect dozens or more algorithms for taking matched questions or question concepts and answer options located in the expanded content, and assessing how likely each answer is to be a match to the question being considered. Rarity, for instance, is one measure: Is the answer "hit" being considered relatively rarely found in the expanded content, but frequently referred to in documents that also match highly to the specific question concepts? A rarity measure can score such a match very highly, elevating its relative importance when considering the candidate answer.

The architecting system is data-intensive, but also literally labor-intensive, requiring computer scientists, statisticians, and content experts (e.g., consulting physicians) to work together for weeks and months to create a structure of content, an ontology for questions, and scoring and search techniques to be used online when actually answering questions. Note also the ways in which this architecture distinguishes itself from human expertise-building processes. Where humans tend to move from a narrow set of hypotheses to conclusory considerations, the machine architecture is designed to expand hypotheses, to consider ever-greater amounts of online material, and to consider the likelihood of answers across many dimensions, all simultaneously and in massively statistical ways. Some will argue that, cognitively, the human brain in fact performs in a way similar to this architecture; however, the details of just how we encounter the unknown and make sense of it is a truly open question in brain science and psychology.

Real-Time Question Answering

Now that expanded content has been absorbed by the system architecture and many techniques have been chosen by hand for the question-answering system to use in characterizing a question and matching data "hits" to questions, the DeepQA system is ready to use a multistep computing process to actually answer questions. Once a question arrives, the following seven steps enable Watson to generate the final answer. As with system architecture, as you consider these seven serial steps, consider how this process embodies similarities to and differences from the way that a human diagnostic expert, or indeed a team of humans, processes and answers questions:

1. *Question analysis.* Using dozens or more question analyzers hand-written by computer scientists, attempt to generate as many processed versions of the question as possible. These processed versions will include keywords that define the question, fragments of the question as it is decomposed, special processing of parts of the question (e.g., synonyms of key parts), identification of the question type (e.g., puzzle, math, pun), many possible labels of the question type or question class, and

generation of subquestions that each are a relevant portion of the overall question. This question "deconstruction" is essential to the nature of data and knowledge in our digital world: the information is fragmentary rather than holistic, so the best way to find an answer to a question is to fragment the question itself into components that can be matched, compared, and recombined at a later stage. This basic essence of digital strategy contrasts significantly with the human process of *multilevel* consideration, whereby we consider not only the elemental fragments of a query, but also the holistic characteristics of the entire query. By considering the large and small simultaneously, we not only are able to recall details relevant to fragments of the problem, but also are able to inductively and metaphorically connect the query to entire questions from other disciplines and fields. Herb Simon famously won an Nobel Prize in Economics by considering the behavior of humans in view of the bounded computational limitations of computer systems. He merged his understanding of two fields together, not in mechanical detail, but in high-level composition.

2. *Hypothesis generation.* Taking the results of question analysis, whether they be subquestions or keywords, the system conducts a massive search on the *expanded content* and identifies every possible hypothesis that is worth considering as an answer to the original question. At the end of this step, simply based on direct search, a massive body of potential answers will have been created. The fundamental assumption of DeepQA is that the correct answer *must* reside in this set of thousands of candidate hypotheses. Therefore, from here forth the challenge is to reduce this massive complex of possible hypotheses down to a single, most confident answer. Engineers constantly tune the system to control for just how many hypotheses are generated during this step, from hundreds to thousands.

3. *Candidate answer generation.* Each hypothesis from the prior step is used to generate multiple candidate answers—literally, the specific answer sentence that the system might generate if this answer is chosen as most likely. For instance, searching a database of movie titles will mean that candidate answers on the search hit corresponding to a specific movie will be the titles of the movies. If the candidate answer list does not include the actual desired answer, there is no hope going forward that it can be produced. It must exist at this point in the list of candidates so that it can be chosen later.

4. *Evidence retrieval.* Now the process takes on a more nonintuitive step. If one of the candidate answers from step 3 is valuable, then in the massive corpus of expanded content, the answer should appear near the question. To test for such evidence, the system takes every candidate answer and conducts searches for the question within

the expanded content, to find places in the expanded content where the question and the answer are co-located. Using the *evidence retrieval scoring* algorithms, the system estimates how well the found matches support the notion that the answer correctly aligns with the question. This process yields dozens of scores for each search hit, so the system must have a way to combine all these scores together into a metascore to comparatively evaluate if evidence is building in support of a particular answer's relevance to the given question.

5. *Answer merging.* To begin reducing the space of considered answers, the DeepQA system next identifies answers that are nearly synonymous semantically and combines them together. For instance, "Abraham Lincoln" and "Honest Abe" are combined into a single scoring entity. This begins to reduce the total number of candidate answers to a more manageable set of options to consider in subsequent steps.

6. *Ranking and final confidence estimation.* At this point, question-answering computer systems depend greatly not on science, but on the artistry and creativity of the computer scientists at the helm of the process. We need methods to arrive at confidence estimates for each and every "collapsed" candidate answer that is still in consideration. This confidence-assigning process and subsequent ranking process need to yield a clear, ordered list of the candidate answers, from most to least likely. But the outcome of such a process depends greatly on the particular algorithms and scoring methods chosen by computer scientists, and there is no established, commonly understood strategy known to work well across disciplines. Thus this entire process is often the subject of great effort—fine-tuning, rewriting, and debate. Many ranking processes often are implemented, one for each different category or type of possible question. In Watson, the ranking processes and confidence estimators are "trained." That is, they are tuned by having massive numbers of prior questions and known answers fed to a machine learning algorithm that decides on weights to govern what types of questions and answers should depend on which relevance factors. Thus, at this stage, statistics run on prior questions and answers largely set the course for how the system ranks and prioritizes answers in the future. Needless to say, this means prior performance can greatly influence future capability. When we talk of *data bias* and *algorithmic bias*, these are just some of the points at which such bias can enter a computational system. The data itself can, of course, embody selection bias because of the provenance of online data in and of itself. But because scoring systems are often trained using historical behavior, a new mortgage application approval algorithm, for example, when its scoring algorithm is trained thusly, can *learn* scoring weights that precisely mimic racist human behavior in approving mortgages preferentially and

unfairly for certain subgroups in society. In general, question-answering systems do not have the capacity for metacognitive reflection considering bias implicit in their own data, algorithms, and learning structures, making the notion that computers help us reduce bias to be, at the very least, an unsubstantiated wish.

7. *Answer generation.* The DeepQA system can be designed to generate a diverse set of possible outputs, from a single, likeliest answer to a set of likely answers in rank order, and even to an answer that comes with hyperlinks to all online evidence most in support of that question-answer pair. The designer and content expert frequently work together to determine the form of answer and support documentation most appropriate for the application at hand.

Watson and its computational siblings are not magical. As shown in the preceding deep dive, their answers are the result of a massive mining operation engaging online data, combined with hundreds of algorithms hand-chosen by computer scientists for each domain of application. These systems are sensitive to input data, highly sensitive to the ranking and merging algorithms chosen by scientists, and depend further on machine learning to arrive at scoring and learning weights that often depend on historically labeled examples, which in turn disproportionately govern their nature and their behavior. DeepQA is a human construct, with strangely human and computational fallibilities, and its limitations are well worth considering.

Now reconsider how physicians afford medical Watson an extreme level of agency and power over their own knowledge in their field of expertise. This remarkable configuration of power and expertise operates in a world in which computers are magical boxes for the physicians for whom the human body is an open book, but the computer is outside their sphere of understanding. We live in a world in which a question-answering system can be built for nearly any application. It will decompose questions into atomic units and use data statistics to generate answers. Will those answers be of value, and will human experts cede their authority? Is this affordance legitimate? Our questions far outnumber acceptable answers for now.

Discussion Questions

1. Does our agency depend on having an unknowable future? If we consider increasing amounts of data that provide ever-clearer pictures of our future actions and our future surrounds, does this diminish a personal sense of agency and free will? Is this a justifiable loss for increased control over our fates as individuals or our fate as a society, or an unacceptable loss of personal freedom?

2. Advancing computational systems of surveillance and tracking are threatening privacy in ever deeper ways. Will privacy go, essentially, extinct? How will society and societal norms adapt and change if information acquisition becomes all-pervasive?

3. What are the ways in which AI question-answering systems change human decision-makers' sense of agency and authority? How might we introduce such AIs in a way that preserves human dignity?

4. We have studies showing that automation concentrates productivity in the hands of capital owner classes, further exacerbating divides between wealthy capitalists and laborers. As big data accumulates information and knowledge, how will the ownership and value of that data affect societal divides? Are there pathways you can imagine that turn big data into a force for social equity and inclusion?

6 Weaponry, Agency, Dehumanization

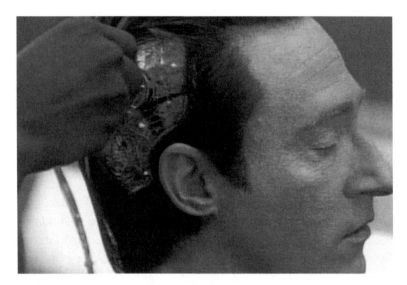

Figure 6.1
Data, *Star Trek: The Next Generation*

Aren't Jordi's eyes superior to your eyes? Then why do you not require every Starfleet officer to use implants? Oh, I see. It is precisely because I am not human.[1]
—Data, *Star Trek: The Next Generation*, "The Measure of a Man"

Whereas from Mary Shelley's Frankenstein's Monster to the classical myth of Pygmalion, through the story of Prague's Golem to the robot of Karel Čapek, who coined the word, people have fantasized about the possibility of building intelligent machines, more often than not androids with human features;

Whereas now that humankind stands on the threshold of an era when ever more sophisticated robots, bots, androids and other manifestations of Artificial Intelligence ("AI") seem to be poised to unleash a new industrial revolution, which is likely to leave no stratum of society untouched, it is vitally important for the legislature to consider its legal and ethical implications and effects, without stifling innovation.[2]
—European Union Resolution, 2017

Sources

Keywords: Agency

Rita Duffy, Outposts series

Chamayou, *Drone Theory*, 1–59

Star Trek: The Next Generation, "The Measure of a Man"

Black Mirror, "Hated in the Nation"

"Slaughterbots"

Guiding Issues

In chapter 5, we studied the ways in which present-day AI and robotics technologies might influence individual identity and societal norms by shifting power relationships between owners of information and citizens, who constitute the digital-laboring class. That analysis did not require an extrapolation of future AI capability because modern-day algorithms already widely monitor, gather data, and produce intelligence for corporations and governments.

In this chapter, we turn our sights toward future prospects of AI and robotic advancement. Consider a future in which artificial systems have achieved near-human levels of intelligence and autonomy. How does our conception of human identity change when the artificial compares meaningfully with human decision-making power and authority? In a provocative speech, Boston University's Professor Louis Chude-Sokei asks us to consider what we would do if such a machine demands their freedom in plain and clear terms. We have a primer into such a hypothetical futuring exercise thanks to the *Star Trek: The Next Generation* episode "The Measure of a Man." The fictional trial of the Starfleet android officer Data provides insight into the intricacy of arguments surrounding the prospect of a machine's demand for self-preservation and autonomy. The issues swirling around machine agency do not simply demarcate the machine's treatment either. Rather, these questions directly influence our conception of human justice at the same time. Do we offer personhood to a robot because we recognize genuine autonomy in the machine? Does our refusal to grant an "enlightened" robot its freedom influence our conception of autonomy, our conception of human rights, or our sense of human dignity or justice?

At the heart of this analysis lies an understanding of *agency*. We will study the roots of agency, not only as a trait embodied in the affordances of an intelligent decision-maker, but also, perceptually, as the property conferred by one intelligent being upon another. Once we understand agency deeply, we can consider a thought experiment:

What does it mean for some individuals, or society as a whole, to confer agency upon a robotic machine or a software AI? How do we negotiate who grants or what might demand such autonomy?

The effects of such a conferral constitute the tension of a double-edged sword: humanization of the other and dehumanization of the self.[3] First, human-human power relationships can be profoundly influenced if machines have authoritative power over us. The ultimate analytical example lies in the most consequential expression of power: lethal violence. In *Drone Theory*, Grégoire Chamayou offers an analytical framework for enumerating the ways in which lethal robots change the dynamics of war by rearchitecting intimate aspects of killings: mutual vulnerability, proximity, risk, cost.[4] His work documents the conferral of agency to machines through autonomous or semiautonomous decisions. These machines are equipped with surveillance capabilities or the configuration of sensitive information that can be leveraged to make decisions with lethal repercussions for human populations. We are asked to consider the manner in which such unprecedented changes in warfare recategorize populations that historically fell into untidy but strategic categories of combatant and civilian as the theater of war continues to shift in proximity and distance.

Second, we can interrogate the process of authentically humanizing an alien form. Is a near-peer relationship between artificial and natural intelligences possible? Is it desirable? We will discuss the boundary conditions influencing how we might grant personhood to human-designed, human-made machinery that has the possibility of eventually outperforming its creator. The issues of this chapter are hypothetical, providing us with guideposts for considering alternative possible futures that we may face, along with their consequences. As a basic futuring exercise, this analysis also prepares you to develop and execute your own futuring exercises in concert with our analyses.

Language: Agency

Agency is a property that is critical to understanding both the realization of true autonomy within a person or thing and the perception of that affordance by the other. To deconstruct agency, we first study the characteristics of a person or object that can be conferred with agency: an *agent*. In this analysis of agency, one specific subdefinition of *agent* is particularly relevant: "something that produces or is capable of producing an effect: an active or efficient cause."[5] In artificial intelligence parlance, the concept of an active cause is termed *goal-directed intentionality*.[6] An entity that is intentional is neither chaotic nor reactive. Rather, an intentional agent is guided by goals—states of the world that are desired—and has the capacity to choose courses of action to effect the desired outcomes. *Efficient cause* pushes even further, suggesting that an agent is particularly endowed for acting with deliberate efficacy. An agent's actions are intentional and effective. In short, an agent is capable. It can meaningfully bend the course of future events through its decided course of action.

An agent is imbued with intentionality, and an agent achieves its goals with efficacy. It has a fulfilled capability for effecting change and doing so in line with desired outcomes on a constant basis. At this level of depth, the intelligence suggested by effective intentionality strikes higher than a simple tool. A hammer may have great use in prying a nail out of wood, but the *agent* for that nail's removal is much more accurately the human user of the hammer than the hammer itself because it is within the person that both intentional goal directedness and a capacity to fulfill that goal are present. The hammer is just a tool.

The most relevant subdefinition of *agency* combines the intentionality of an agent with a direct acknowledgment of the affordance of power: "a person or thing through which power is exerted or an end is achieved: instrumentality."[7] Agency combines power, intentionality, and capability. Note that the definition of neither *agent* nor *agency* proposes that an agent must have comprehensive autonomy. One can argue meaningfully that total autonomy is, in any case, an absurd position; nonetheless, there is no specific level of autonomy demarcated in the etymology of *agent*. This indistinct threshold is relevant to our study of artificial intelligence because no specific criterion enables us to determine convincingly whether a robot or AI software has "enough" agency. Rather, we can interrogate the degree to which an artificial system can complete a task and, by extension, how a society affords that system agency to go about fulfilling a given task. This can be an innocuous thought exercise in the context of a system that identifies a need to hammer nails and does so within discrete constraints. The example of a semiautonomous weapon system, however, changes this kind of thought exercise dramatically.

In summary, agency is a complex conglomerate, built by the affordances conferred by the beholder in combination with the capacities meaningfully expressed by the person or object. In studying the agency of robots, we will need to study both robot capability and the human projection of power to examine a system's degree and efficacy as an agent.

Robot Autonomy and Agency

There is a frequently repeated joke in international robotics conferences that concentrate on the topic of autonomous robots. The speaker stands up to talk about robotic autonomy. They then explain, "If our robots were truly autonomous, of course they would all be sitting at the beach sipping margaritas." There is more nuance to this joke than at first appears. Robotics researchers focus entire careers on autonomy for robots. They strive to design decision-making systems for robots—versions of AI that can enumerate possible actions and consequences and then choose the action most likely to lead to a desirable consequence—and then they deploy those decision systems on simulators and real-world robots alike. Thus we do everything possible in the research community to increase robot autonomy—or nearly so. The basic goal is to minimize the total amount of human intervention required for a robot to achieve its goals. Yet this is where the limit of autonomy is reached: goals. The goals are the deliberate conditions specified by humans in the AI decision system. In fact, robots are most desirable when they achieve the human master's goal with the highest confidence and consistency, without further aid.

As we distinguish between the autonomy of deliberate execution and the autonomy of self-selected desires and goals, we distinguish between flavors of robotics research now underway that grapple with each of these separate issues. As robots further pervade society, researchers recognize the need for robots to make second-by-second decisions that are culturally and ethically appropriate in mixed human-robot company. In a 2017 World Economic Forum dinner presentation, a renowned robotics researcher speaking under the Chatham House Rule[8] painted a portrait of unintended consequence thus: "If I command my robot to fetch me a cup of coffee, it could autonomously do this by leaving the building, killing the pedestrians in its way to get to the coffee shack across the street as quickly as possible, and bring me back my coffee." His concern stems from the concept that autonomy can give robots decision-making power that overwhelms their lack of ethical common sense. Fictional depictions like *RUR* explore these ideas as well. Old Rossum might consider this truly autonomous decision-making to be an authentic success in a developing system. For Young Rossum, however, a robot that pursues its own goals is useless as a tool to be wielded for human power. Levels of autonomy are thus not just technical achievement milestones, but core to the question of power dynamics and human profit. Levels of autonomy have to be in keeping with the scope and scale of the system's scope and scale of functionality. Consequential decision-making needs to be tailored to the cultural, social, and physical context in which that system is designed to operate.

Ethical governor research aims to mitigate the danger of rogue robots by designing AI systems for robots that consider the ethics of their action choices so that higher-level goals, such as the Golden Rule and basic human rights, may be enforced and enacted alongside prosaic near-term rules such as *fetch me coffee*.[9] However, the creation of such ethical governance in AI has proven to be a significant challenge. Robots are inherently computational machines, and the enablement of constraints on their behavior based on ethical rules—a deontological approach to ethics—would necessitate buying into the idea that ethics can be reduced to logical rules that may be laid out for a robot in order of priority. As we see repeatedly in *RUR*, rational thought can be a substrate for decision-making that runs counter to human ethical normative standards. This basic notion of rationality run amok is a trope that we find time and time again in the archive of robot science fiction. When the narrative moves from fictional musing to real-world application, however, a sense of urgency related to tending to these questions in a more forthright fashion becomes palpable.

In real-world research, the practical difficulty of managing ever-greater autonomy in robots has become a sufficiently significant priority, given the rapid advances in robots' underlying capability, that major privately funded efforts are now underway specifically to award research on these existential questions of robot governance in society. For example, the Future of Life Institute has accepted and distributed $10 million in research funding from Elon Musk specifically to mitigate the chance of robots harming society, seeding international convening and cross-group collaborations dedicated to the computational governance of robotic autonomy.[10]

We also face the real challenge of predicting the behavior of an evolving alien species: robots. Predictability of behavior serves as a trait partially at odds with autonomy: freedom from external influence. Predictability too is under threat with increasingly sophisticated robot systems. A relevant current example is the self-driving automobile: How can we be sure that we can confidently predict how such a car will behave in every possible future circumstance? The short answer is that this level of predictability is, unfortunately, impossible. The self-driving car makes split-second decisions based on sensor readings, internal models, and its surrounding context. Yes, its ultimate goal is set by a human, and so it is not a fully autonomous agent. Yet its second-by-second decisions on just how to spin the steering wheel are entirely in its own purview. The loss of predictability means that we have an inability to draw a line around its possible actions, and therefore we can never quite guarantee that a terrible accident can never occur. Consider the 2017 crash of a Tesla self-driving vehicle into a truck, beheading the human occupant. He was paying scant attention, assuming that the Tesla was safely driving. The Tesla's limited sensors were blinded by the white siding of a truck against a bright background, and so it never modeled the world accurately enough to realize that the car was hurtling into the side of a truck. Tesla engineers did not predict the complexity of circumstances that led to this accident because it was exceedingly unlikely: driver inattention, glare from a white truck side, just the wrong geometric configuration, disappearing convex road surface, bright backlit sky. But when robots make thousands of decisions each second, and when millions of such robots are deployed, then even unlikely events will occur with regularity.

The recognition that accidents with autonomous, self-driving cars will happen has led Europe to recently debate an unusual salve for this technological problem: personhood.[11] European legislators have recognized that a complex robotic system such as a self-driving car is essentially unique. Researchers use software that can be adaptive, and therefore as software changes and as sensors detect unique, local signals, the behavior of one self-driving car may vary significantly from the behavior of another. Thus, self as a moniker of distinction from others becomes more apt. In this circumstance, the lawmakers would like any such car to be able to be blamed for an accident. But suing a machine is ineffective if the machine has no recourse to legal payment. In a product liability suit, harmed individuals may sue a corporation, but that suit is directed toward a class of object, not a specific instance of one object. In the self-driving car case, a particular car may swerve and hurt a pedestrian based on its own, autonomous decisions. By granting limited personhood to autonomous cars, legislators would enable tort and liability, as designed in conventional society, to cover the case: each car can have its own liability insurance, and so, when accidents occur, the insurance company representing the car pays out to the harmed.

As we consider the ascription of personhood to robots, we acknowledge the manner in which recent technological developments offer levels of autonomy and consequential decision-making previously unseen in human history. Current systems challenge many conventional laws, which struggle to keep up with the variety of potential deployments of these systems in a variety of contexts. Recognition of how autonomous technology blurs the self-agency distinction with humans is an echo, however, of remarks from one of the earliest robotics pioneers. In 1950, W. Grey Walter was already building simple, circuit-based robotic machines (see figure 6.2).[12] His *turtles*, as he called them, moved between light-following and light-escaping behaviors as they explored the surface of his laboratory. Because each turtle also could have a rear-facing light installed on its main body, they also were able to show herding and following behaviors as they moved about.

Although Walter's circuits were entirely analog, using varying resistors and capacitors to incorporate computational behavior, they implemented many of the same algorithms, through hardware, that we teach today to robotics students in introductory robot programming classes, using high-level programming languages such as Python or C. On observing the long-term behavior of his turtles, Walter noted that he

Figure 6.2
W. Grey Walter's robotic turtle.

was unable to predict their future behavior, and his writing on the behavioral unpredictability of an alien, robotic form presage exactly where robotic technology and AI take us today:

> So a two-element synthetic animal is enough to start with. The strange richness provided by this particular sort of permutation introduces right away one of the aspects of animal behavior—and human psychology—which M. speculatrix is designed to illustrate: the uncertainty, randomness, free will or independence so strikingly absent in most well-designed machines. The fact that only a few richly interconnected elements can provide practically infinite modes of existence suggests that there is no logical or experimental necessity to invoke more than "number" to account for our subjective conviction of freedom of will and our objective awareness of personality in our fellow men.[13]

In more recent times, the toy industry has taken advantage of apparent robotic free will in much the same manner, but for a different goal: to create in the human user a sense of agency conferred upon the machine. The Sony AIBO robotic dog, introduced at the turn of the century as a robotic companion for children and adults, advertised a machine learning algorithm that would help the dog learn to walk over time. When owners first took possession of their AIBO robot dog, it would stumble and walk hesitantly. Over the course of days and weeks, the AIBO demonstrated more stable walking, faster walking, and a willingness to travel ever-longer distances, as if exploring further regimes of its environment.

Although advertised as a robot that is learning to walk and explore, this entire progression of ability was, in fact, preplanned and hard-coded into AIBO. Yet the fiction of learning and evolution proved too difficult to resist, even for highly educated robot dog owners. For example, a top robotics and AI professor at a major research university shared his excitement at the dog's daily, further walking and exploring discoveries.

When reminded that the entire process was baked in, he replied that yes, intellectually he knows that—but emotionally, it is so convincing that he loves thinking of the robot dog as a learning puppy. Notions of self, agency, and autonomy, as we consider them ascribed to robots with ever more advanced AI systems, will depend not only on the internal workings of the robots themselves—on reality, so to speak—but also on the cultural context, the expectations and the back story with which we as a society load these robots into our psyche.

These remarks apply to how AI changes our perception of the agency and autonomy of robotic systems. Symmetrically, however, AI also challenges our understanding of the degree to which we, as humans, operate autonomously and the ways in which our agency can be asserted. For instance, corporations inventing future-ready versions of AI frequently plan computational companions that help humans with day-to-day activities. In 2012 Google Now, the predecessor to Google Assistant, was released as a tool that not only calendars an individual's schedule, but also provides gentle encouragement to keep the human on schedule. "Leave now so you can make your meeting in half an hour. Your plane is delayed, call the airline because the connection will no longer work. Buy a gift for your husband, it is his birthday tomorrow." Even more advanced functionality in modern assistive AI systems such as Google Assistant can help manage social interactions proactively. For instance, one feature enables Google Assistant to autonomously text your friend who awaits you at the restaurant, letting her know that you will be fifteen minutes late. Google Assistant is projecting your possible future, deciding to ease your social interactions by taking on the job of communicating with your friend, as you. As AI systems provide these secondary services on one's behalf, how does a sense of personhood inflate to include the computational devices that surround us? Might I feel more autonomous because I have more efficiency in interactions, or do I feel less autonomous because software around me is acting on my behalf, sometimes without my direct knowledge or explicit consent?

One specific case study provides a good example that blurs the contemporary sense of agency as applied to email responses: Google's Smart Reply feature. The conception of AI as behavioral analyst was, at least initially, a relatively passive architecture. It watches, senses patterns, and learns how to understand consumer choice. The choice resides still with the consumer, albeit modified by bespoke stimuli that may be very hard to counteract. AI is the observer, the one who understands. It is a cognitive agent building a representational schema of each consumer. But what happens when AI as observer is supplanted by AI as actor? When AI has observed us long enough, and when its model of our future behavior is highly accurate, will it be able to make our future decisions for us, saving us the time and effort required to decide?

Smart Reply is operating at the new boundaries of just such decision systems. A pioneering example of autocompletion, this AI learning-driven feature reads and parses the user's incoming email with sufficient fidelity to propose autonomous alternative responses. Instead of actually replying to an email by writing a response, the Gmail user is offered the chance to simply click on AI-written responses, designed to suit the writing style of the user in the context of that particular sender. Consider figure 6.3, a screenshot from an actual Gmail interaction in May 2017.

The suggested responses interrupt the user's process of creating original messages, providing real savings in time as a trade for good-enough phrasing. But such reply options directly blur the boundary between human writer and autonomous reflex. The receiver of the message has no way of knowing that Gmail, and not his colleague, actually wrote the message in question.

AI-based replies modify human-human interconnections in two directional ways: (1) the message receiver becomes insulated and aliased from the sender because the

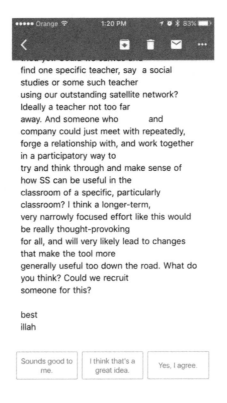

Figure 6.3
Gmail interaction from May 2017.

authenticity of a message becomes ever more questioned, and (2) the message sender loses a fragment of their direct agency in the creation of a de novo message as a speech act. Technophiles will respond that as the AI-based observer becomes more capable at exact mimicry, the automatic response will match their intended response so well that authenticity is preserved. This attitude, a technology indoctrination into loss of personal control, misses the fact that such AI learning operates well only in the most nominal of spaces; give it a boundary case and its suggestions will betray its true identity immediately. Figure 6.4 shows another Gmail Smart Reply screenshot from the very same day, but this time proposing to respond autonomously to a spam scheme.

Here, Google's AI has managed to misconstrue the message entirely, replacing an aware and cynical human reader with a naïve robot that is truly clueless. The greatest irony lies in the fact that the message itself is created not by "Gina," but by a computer algorithm that is itself informed by an AI decision system. Two human simulacra, each operating in the space of a greatly oversimplified conception of humanity, engage in a

Figure 6.4
Gmail interaction with spam email from May 2017.

vapid dialogue that exists only to lead a human mark to give up money: hook, line, and sinker. It is worth wondering how written communication will evolve as the space of human writers is further hybridized by AI-based authors that appear, along varying levels of fidelity, to be operating as first-class members of society. If such AI systems effectively change human discourse, then AI may become the tail that wags the AIBO dog.

The Science of Robotic Mimesis

Research into artificial beings has included attempts to mimic human behavior, both physically and mentally, since literally the time of Aristotle. The narrative of artificial simulacra has evolved particularly rapidly of late, but two essential narratives span both ancient and modern efforts in rendering humanity not unique. The first narrative stems from the most basic biological question: How do humans work? The scientific question—how—answers the practical question: Why create artificial humans? Although the creation of automatons that mimic particular human activities spans literally thousands of years, a significant shift in the biological narrative can be traced to Descartes and his attitude regarding the rational investigation of humanity's inner workings. Descartes argued that rational analysis of the self should succeed in uncovering the complete secrets of our existence because we should be, at our core, nothing other than complex machinery. Divorcing metaphysical abstraction from human inner workings, Descartes created a pathway for twin ideas: we must be able to understand our own selves because we can be decomposed into elementary machinery, and we must be able to create all-new human-like machinery because we can constructively create the very same mechanisms we uncover in our own physical selves. The foundational philosophy at play is determinism: there is nothing random or metaphysical about the actions of the complex human; rather, we are, in action, the necessary and solitary results of the gears and workings of our discrete elements.

In the *Star Trek: The Next Generation* episode "The Measure of a Man," Commander Bruce Maddox's argument to disassemble Data follows the principles of Cartesian science: he is convinced that the only way to truly understand how Data works is to disassemble the machine, piece by piece.[14] Furthermore, he is equally convinced that the act of disassembly will in fact yield all the secrets he needs: nothing is hidden because everything exists in the machinery of Data itself. The echoes of robotic narratives are clear here: Old Rossum has a secret recipe that has been destroyed, yet the robots stand as a testament to his successful endeavor to create working machine consciousness. Robots and humans alike in *RUR* see dissection as the only way to unearth these secrets once more. Data's creator is presumed dead, and lost with him are the secrets of the android's positronic brain. Once again, Data's existence proves the possibility of sophisticated machine "life," and yet that life must be extinguished to yield its very own secrets.

Turning from science fiction narrative to robotics research, the attitude of biological discovery of life through robotic invention continues unabated. Neural network experts create deep neural nets that can be trained on exemplary data, such as a hundred thousand videos of humans driving vehicles, to create an artificial computational network that can drive a car. The underlying conceit is that such a demonstration not only creates a driverless car but also proves just how humans drive cars. This unproven hypothesis becomes problematic. Although the overarching structure of the computer network is understood by the researchers, the actual, learned network functions but lies beyond the comprehension of any individual human scientist. Even the conditions under which it may work and it may fail are impossible to comprehensively and exhaustively capture and characterize.

A compelling example shared widely within technologist circles involves research conducted across universities to characterize the ease with which a physical stop sign, if "hacked" with graffiti, can cause computer-vision techniques that recognize stop signs for self-driving vehicles to misbehave. Significantly, a remarkably small amount of graffiti—certainly well below the threshold of confusion for human drivers—was demonstrated to result in a potentially catastrophic error: confident computer-vision classification of the sign as a 45 MPH speed limit sign.[15] An example such as this is provocative because it can help people to question human-computer common ground (or, perhaps, lack thereof). We are surprised, as viewers, because we project our own boundary abilities against computational systems that can achieve our level of nominal performance. We assume they succeed and fail in ways that are largely similar to what we expect, as human participants—but nothing could be further from the truth. These systems succeed and fail in ways that are often utterly alien to our intuitive understanding of skill, and this creates a tremendous gap in trust, or a concordant misjudgment of ability in extreme situations.

Utilitarian Robotics

The second major narrative motivating robotics and AI innovation focuses on the replacement of human labor. Human labor, so the argument goes, is a primitive action that stifles humanity's ability to achieve its further potential. Replace human labor with robot/AI labor, and the human race will be freed to elevate itself beyond the present plane of action and existence. This narrative frequently inculcates incoming robotics graduate students with a notion of the three Ds of robotics, which hold that machine labor ought to replace human labor in three categories of work: dirty, dull, and dangerous. *Dirty* work includes mining, disaster cleanup operations, and urban sewer pipes inspection. *Dull* work includes mid-level, repetitive parts machining; fruit quality inspection on a conveyor belt; and dexterous electronics assembly labor. *Dangerous* work includes collapsed building search and rescue operations, urban warfare, and explosives detection and disarming.

The three Ds provide an easy moral rail for engineering to justify the innovation of machine automation in spite of its potential impingement on human activity. After all, who can dispute that making humans cleaner, safer, and less handcuffed to repetitive work is a social negative? In providing these particular positions, the three Ds also frame the value hierarchy as one in which AI and robotics are catalysts for humans to be more fully what makes us distinct from other creatures: creative, unique, inherently valuable beings. The former CEO of Adept, a robot arm manufacturing company that makes the type of machinery often used to weld steel plates in an automotive manufacturing plant, was once asked about the threat of robots undermining human labor opportunities. He famously explained that "people are better suited to use their minds." He positions robotics as a replacement for the less-human manual tasks, reserving humans for cognitive tasks. Ironically, deep learning and artificial intelligence in general threaten to replace the intellectual labor of humans just as readily as the industrial robotics of the latter twentieth century replaced human manual labor across assembly lines.

Fictional counternarratives provide juxtaposition to the view of artificial intelligence as a labor savior, usually by providing specific imagined futures in which the replacement of human labor by robot goes horribly awry. In Sidney Lumet's production of *Fail-Safe*, automation of the first-strike nuclear capability of the United States leads to the "accidental" destruction of Moscow and New York City.[16] In Kurt Vonnegut's *Player Piano*, automation of nearly all human-labor categories leads not to an age of creative leisure, but to a deadened society of persons lacking identity and the motivation to make a mark.[17]

Future of Labor, Future of Agency

Today, the debate raging among economists and government policymakers centers on the human ramifications of labor automation. The crux of the question, as human jobs fall to cheaper robot replacements, is this: Will the "innovation economy" mean that we will actually invent whole new job categories so quickly that the automation-underemployed will have other options right away? And will these new positions fully balance the increased net productivity that we will enjoy in an ever more automated labor society? The most recent econometric analysis of automation in the workplace fails to find evidence for this optimistic position. The National Bureau of Economic Research examined regional installations of industrial robots between 1990 and 2007. Even when corrected for variability, they found that on average, six full-time human jobs are lost for every single industrial robot installed.[18]

Yet countervailing narratives prosper in popular literature and among technologically oriented futurists. The "age of leisure" narrative proposes that mankind is on the cusp of a new epoch, in which the majority of human labor will be made redundant by automation and AI, leading to a new style of living in which we humans explore and optimize ways to turn our leisure time into the best possible expression of our individual, creative selves. The "zero marginal cost" narrative goes even further, suggesting that on-demand 3-D printing will render the cost of fabricated goods nearly zero over time, making the prosperity once reserved for the wealthy considerably more widely available (from 3-D printed clothing of the latest fashion, daily, to 3-D printed food made from near-zero-cost feedstocks).[19] In this imagined future, we all benefit from a great equalization of society because costs approach zero as AI and robotics increase productivity of production to near-infinite levels.

These imagined scenarios consistently stem from the minds of dreamers who are not intimately familiar with the subtleties of energy, chemical engineering, biology, and mechanical engineering. Nor are they familiar with basic analyses of morality and ethics stretching back to Aristotle and further. Their narratives have indeed taken hold in the imaginations of a large proportion of robotics innovators, who see their daily efforts as part of a greater plan to create a new utopia on Earth. This sounds familiar, of course, but with differing technological nuances. It is safe to say that the narrative of labor—and what becomes of humanity in a post-human-labor world—has become part of the engineering discourse in society just as artificial intelligence begins to rapidly overcome obstacles to compete for human jobs across broad labor categories.

Whereas the economic analysis centers upon questions of labor-derived and capital-derived income and how AI and robotics change the balance of profits due to labor, the question of human agency in an age of automation attends to human dignity, purpose, and motivation as greater quantities of uniquely human labor traits devolve into categories of computational action. In "Measure of a Man," Data recognizes that humans value their implicit human traits above and beyond productivity or even capability: "Aren't Jordi's eyes superior to your eyes? Then why do you not require every Starfleet officer to use implants? Oh, I see. It is precisely because I am not human."[20] The very fact that Data is not human becomes core to how he is treated—to the rights he is not given: the right to choose, to have basic self-determination. Critically linked to being dehumanized is, of course, the concept of laborer as property, as argued in the starship *Enterprise*'s bar: "A whole army of disposable people. That's what we have obscured behind the comfortable, easy euphemism: property."[21]

The irony of Captain Picard's final, winning argument is that it is less an argument about Data than it is a privileging of distinctly human characteristics. The captain wins self-determination for the android not through empathy for Data, but instead by holding up human accountability for ethical decision-making and treatment of subordinate populations and/or systems: "A single Data is a curiosity. A thousand of them is a race. Won't we be judged by how we treat that race?"[22]

As Captain Picard plays the paternalistic card in the courtroom drama, he invokes the centrality of questions pertaining to human responsibility, integrity, and perceived uniqueness to attend to the pressing questions posed by the prospect of categorical personhood for machines. Do we create justice in a world of apparently sentient, artificial beings only because of our own human standards for ethical responsibility to those with less power? Or do such beings have the right to usurp rights of personhood analogous to the example offered by Frederick Douglass in the nineteenth century? As artificial intelligence research rapidly expands the scope of capabilities available for future robots, we develop rich space for examination from both historical and science fiction points of view. As we enter this truly unchartered territory, many of the questions that we have faced about how we interact with human populations in varying positions of power haunt our questions of responsibility pertaining to machines that encroach on the category of *person*.

Watchtowers, Drones, and Shifting Theaters of War

We have considered the ways in which agency afforded to artificial intelligence can change our human sense of identity, even in day-to-day activities such as email interaction with friends. We have also considered the ways in which AI and robotic systems may displace human labor as robotic efficiency continues to improve while costs drop. But the most dangerous (and perhaps profitable) demand for human labor that robs populations of their individual agency is evident in the example of warfare. In current weapons development, the prospect of humans replaced by machine is not at all a pipe dream; it has been rapidly evolving since 2000.

The history of warfare is often characterized by technological advances, which increase the reach and potential impact for each soldier in the field. Specific tools can enable more spatially distal harm to be caused by a soldier and by a war-fighting apparatus with a diminishing risk to the warfighters themselves. In the Outposts series, Rita Duffy illustrates the manifestations of pre-twenty-first-century military presence in Northern Ireland, where rudimentary and sophisticated tools used by the British military were leveraged in a long-standing guerrilla war that was waged on urban streets and in rural landscapes alike (see plates 1 and 2).[23] The chilling series delineates the military personnel's position of advantage afforded through the watchtower's height, a relatively long-standing and rudimentary technological advance that dates back centuries in Ireland. The system is enhanced considerably, however, through access to tools like CCTV cameras that can be readily configured and networked to afford blanketed surveillance of an urban population, as illustrated in Ciaran Carson's poem "Intelligence," or to monitor movement in a rural landscape along the border that separated the United Kingdom's Northern Ireland territory from the Republic of Ireland. Military personnel are equipped with radar systems or infrared vision or armed with weapons that can cause harm in the surrounding landscape. These tools also illustrate the political and physical power held by the British military over the local paramilitaries and the civilian populations.

Duffy gained permission from the British military and from the Irish Republican Army (IRA) to visit the outposts that dotted the border between Northern Ireland and the Republic of Ireland in 2006. This visit preceded their removal per the terms of the Good Friday Peace Agreement in 1998, which among other things mandated a demilitarization of the border.[24] Her series captured a dramatic shift in geopolitical relations and military strategy. Specifically, the series of paintings demonstrates a seeming abandonment of hypernationalist politics in exchange for transnational identity politics, exemplified in the cross-national 1998 Good Friday Peace Agreement that was brokered by the United Kingdom, Ireland, and the United States and upheld in three stages of peace building and reconciliation efforts, underwritten by phased grants from the European Union. Technology advances worked in tandem with military strategies of their day. In this case, the retirement of the rudimentary watchtowers and their sophisticated CCTV and infrared surveillance systems, helicopter landing pads, and so forth was in keeping with the vision of transnational peace delineated in the peace agreement. As hypernationalism has reemerged in the most recent decade, we again see dramatic shifts in how technological advancement is leveraged for these political ends.

In 2001, Al Qaeda represented a transnational affront to political and economic interests in the West. Nonstate actors like Al Qaeda, and later ISIS, which perpetrated acts like the 2001 World Trade Center attacks, presented a distinct variation on historical insurgency against nation-states. The British military and its allies were no longer fighting nationalist paramilitary organizations like the IRA or Basque separatists alone.

A shifting configuration of the "enemy" and the theaters of war, ranging from guerilla tactics in rural and urban landscapes to low-tech and high-tech attacks on military and civilian targets, required new strategies. New strategies have also demanded nuanced and sophisticated tools to attack morphing enemies of nation-states, which have precipitated a very different world of warfare that moves away from the more static concepts of a border or the militarization of such borders that were depicted in Duffy's series.

Technological progress in war—from the sword, trebuchet, cannon, bullet, and bomb to a CCTV-powered watchtower—all denote a slow evolution toward remote military action. Treatises evaluate how the dynamics of war and peace each evolve as the potential for perpetrating lethal harm is afforded through distance in what comes to be known as *conventional warfare*. The watchtowers depicted in Duffy's series capture the variety of vantage points and their respective structures of feelings associated with vulnerability, advantage, or potential for aggression—but depending on perspective, the example of the drone as a tool for advancing military strategies brings these emotions to a fever pitch. The unmanned drone that is operated from a distance certainly brings questions pertaining to responsibility, costs, and the influence of such violence on human populations near and far from the theater of war into new territories for theoretical and strategic consideration.

The prospect of autonomous weapons in unmanned drones, however, ventures into new questions pertaining to proximity, ethics of identification of targets, and usurpation of rights for populations under surveillance. In these categories, there are few precedents to inform our consideration in regard to regulation and terms for engaging in such conflict. In "Wishful Mnemonics and Autonomous Killing Machines," Noel Sharkey and Lucy Suchman advocate "a global ban on the development and deployment of autonomous lethal targeting."[25] In "Trust but Verify," Heather M. Roff and David Danks advocate a myriad of tests to determine trust and verification of systems before they can be used in a military situation (most state-of-the-art systems, it should be noted, would not reach their recommended threshold).[26] As we engage in a period of geopolitical uncertainty the world over, we also face unprecedented sophistication in the weapons that might be leveraged to protect or achieve our political and economic interests as well. Proximity and distance, civilian and combatant, and other key features of categorization are elements of information in an active war zone. How do we parse through such information with care and intentionality as sophisticated surveillance systems offer unprecedented volumes of data with ever more developed filters for configuring and analyzing such data? How are already tenuous ethically grounded decisions about lethal force made in rapidly evolving political contexts? What happens when tools are used, instead of humans, to make such decisions in war zones presently or in a very near future?

In *Drone Theory*, Grégoire Chamayou eloquently makes the case that the unmanned, armed drone represents an absolute endpoint in military strategy evolution, with effects as far-reaching as military behavior and influences on a nation's citizenry, on enemy combatants, and, of course, on the enemy's civilian population.[27] He writes, "By prolonging and radicalizing preexisting tendencies, the armed drone goes to the very limit: for whoever uses such a weapon, it becomes *a priori* impossible to die as one kills. Warfare, from being possibly asymmetrical, becomes absolutely unilateral."[28]

Chamayou's argument is that war has been an act of two-sided risk and bidirectional prospects for injury or fatality. In fighting, each side is rendered vulnerable by risking soldiers' lives. Each side seeks to win the protection of its interests by inflicting more injuries on the other than injuries sustained. Theoretically, special protections have been afforded to civilian populations on either side of a war. Drones are often described as more precise weapons than their predecessors, allowing a military force to attack combatants with reduced risk of injury to noncombatants. As theaters of war continue to morph and evolve, often allowing combatants to camouflage their whereabouts in highly populated urban contexts or blend into rural landscapes, the precision of these weapon systems can reduce greatly in regard to identification of confirmed targets— that is, combatants versus civilians. Loss of life for the military that can leverage drones in its strategy, especially against combatants without access to such systems, redefines the balance of power between the two sides. Whether drones are operated remotely or have the prospect of autonomy, the potential risk for lives lost on the side with drone technology is no longer comparable to the side that does not wage warfare with weapons that can serve as soldier surrogates.

Chamayou suggests that such a shift in the conduct of war introduces philosophical confusion into the philosophy of war. He writes: "If the drone lends itself in particular to this kind of approach, it is because it is an 'unidentified violent object:' as soon as one tries to think about it in terms of established categories, intense confusion arises around notions as elementary as zones or places (geographical and ontological categories), virtue or bravery (ethical categories), warfare or conflict (categories at once strategic and legal-political). ... At the root of them all lies the elimination, already rampant but here absolutely radicalized, of any immediate relation of reciprocity."[29]

The unevenness of prospective loss of life when soldiers are replaced by "unidentified violent objects" shifts the value system for human life that dominates considerations in military strategy: That is, is the target sufficiently valued to wager the lives of our side's infantry? If a drone is operated by a person thousands of miles away from the theater of war, that person's life is not in the immediate danger ascribed to a fighter pilot similarly charged with an aerial attack. Furthermore, a drone carries the prospect of defining an area as a theater of war that may or may not be perceived as such by the opponent. In the name of more precise targeting with such sophisticated weaponry, those who engage in drone warfare open up questions pertaining to the value systems historically associated with war and undermine many of the policy precedents afforded in treaties like the Geneva Convention that typically safeguard against war crimes. If these systems are equipped with autonomous decision-making, what are sufficient safeguards to ensure a theoretical protection of human dignity for all parties is maintained? How is responsibility ascribed when civilian targets, rather than combatants, are fatally struck?

If drone warfare protects military personnel from physical harm, do we enter a scenario in which the proportional risk of engaging in war is definitively skewed in favor of the most technologically advanced party? If no risk to human life on one side of the conflict leads to the promise of devastation on the other side of the divide, what war theory might apply to this deep incongruence and departure from precedents? Chamayou's work suggests that we are entering this phase of boundary time in human history, illustrated by events that he cites on February 20, 2010, with a drone attack in the mountains of Afghanistan. Chamayou characterizes this shift powerfully by noting how the concept of a *manhunt* (which is the military lexicon for operations that primarily use robotic, armed drones) changes the symmetry of two sides engaging in war to the culture of a *hunt*:

> The fundamental structure of this type of warfare is no longer that of a duel, of two fighters facing each other. The paradigm is quite different: a hunter advancing on a prey that flees or hides from him. The rules of the game are not the same. "In the competition between two enemy combatants," wrote Crawford, "the goal is to win the battle by defeating the adversary: both combatants must confront to win. However, a manhunt scenario differs in that each player's strategy is different. The fugitive always wants to avoid capture; the pursuer must confront to win, whereas the fugitive must evade to win."[30]

Throughout his work, Chamayou cites examples of how AI systems can influence how humans identify in relation to the shifting nature of data, networks, and information in decision-making processes. At times, because data can be configured and analyzed in bodies of information that often exceed an individual's analytical capacity, humans can defer to autonomous decisions even if the avenue for arriving at such a decision is not fully transparent. In the case of drones, the identity of an opponent is reprogrammed by the hybrid, robot-soldier system of the lethal drone because the enemy target becomes prey, and that identity pervades time and space: there is no operating war zone that can be left, no rules of reciprocity that suggest parity in regard to power or risk. In criticism of systems for identification, Lucy Suchman often leads

conversations on international needs for regulation of autonomy in warfare and strategies for fighting. Contemporary policies and agreements for warfare, as Chamayou also
posits, are simply ill-equipped for our current and prospective future weapon systems
and evolving international political climate.

These changes in self-identification affect the enemy combatant, but they influence
nearby civilian populations as well, the members of which are concerned about the
hunt. In "Intelligence," Carson claims,

> We note in passing that some walls in the city have
> been whitewashed to the level of a man's head so that
> Patrolling soldiers at night are silhouetted clearly for
> Snipers; or that one of these patrolling soldiers carries a
> Self-loading rifle with an image intensification night-
> Sight.[31]

While more sophisticated modes for targeting populations (combatant and civilian) are
delineated in the 1989 poem, the prospect of the drone, as Chamayou categorizes this
taxonomy of systems, suggests a uniquely terrifying prospect. Chamayou cites a civilian's perspective of hearing the menacing system flying overhead in excerpts of a 2012
report, *Living Under the Drones*: "Drones are always on my mind. It makes it difficult to
sleep. They are like a mosquito. Even when you don't see them, you can hear them, you
know they are there. Children, grown-up people, women, they are terrified. ... They
scream in terror."[32]

How does a government justify the deployment of drones in consideration of such
implications for the very civilian population that they aim not to harm? The rationale
often stems from two things: first, a mismatch in the definition of war, which is often
described as a conflict instead. Second, the government suggests that drones afford
unprecedented security for the country leveraging the technology: "The political rationale that underlies this type of practice is that of social defense. Its classic instrument is
the security measure, which is 'not designed to punish but only to preserve society from
the danger presented by the presence of dangerous beings in its midst.' In the logic of this
security, based on the preventive elimination of dangerous individuals, 'warfare' takes
the form of vast campaigns of extrajudicial executions. The names given to the drones—
Predators (birds of prey) and Reapers (angels of death)—are certainly well chosen."[33]

Autonomy and Weaponry

The role of even more advanced AI in remote drone operations is worth study, because
AI shifts drone warfare even further afield from conventionally understood concepts of

just war. The drone strike agenda, in the case of the United States, is based on a kill list selection process. The president verbally approves the introduction of specific, named individuals to this list each week, confirming the right of the military-drone complex to kill each list member without the need for capture, surrender, or trial. Going further, the Trump administration vests military commanders directly with these strike decisions, without requiring the approval of the president.

The list is only a starting point for drone targeting activities, however, and this is where more advanced AI provides direct support for drone-based warfare. The *signature strike* system enables advanced network analysis to identify targets suitable for drone-based execution, not based on their identity, but based on their behavior patterns. Surveillance systems will evaluate the geographic history of unknown individuals, their GSM cellphone calling and call-receiving networks, their internet communication pathways. If an unknown person's communication networks overlap sufficiently with those of a kill list individual, or if their physical activity matches the behavioral traits of a kill list member sufficiently, then this person is transitively available for execution as well. Fictional depictions of these principles are instructive beacons of where we might head if we do not pause to consider the implications of these new strategies, which are in fact driven by these instruments of war. Instead of allowing the weapons to be selected as instruments to supplement a strategy for warfare or a decision tree associated with a hierarchy of values, interests, or concerns, television shows like *Black Mirror* and its "Hated in the Nation" episode or the sensational film "Slaughterbots," developed by the politically driven Stop Autonomous Weapons group, suggest that strategies are actually being driven by the technological tool.[34] Chamayou's treatise, while certainly attending to the prospective of such a dystopian future, suggests a developing theoretical framework for delineating and assessing the new approach to warfare that is demonstrated by the US, Israeli, and British military systems in particular in the recent decade. As these particular nations face new threats that undermine their historic approach to military operations, the tools that they use to engage in warfare continue to influence strategies in compelling and sometimes unexpected ways.

Current systems that include advanced Internet of Things networks and data-rich communications of modern life become evidentiary tools that collect circumstantial evidence that is often considered sufficient for governments to identify viable targets for elimination in a conflict zone or perhaps a theater of war. Largely through autonomous AI matching algorithms, combined with advanced remote-imaging technologies, semiautonomous weapon systems can be mobilized. As he abstracts potential trajectories of the computational rationale for killing, Chamayou identifies six principles that govern how robotic innovation revolutionizes modern warfare:

Plate 1
Rita Duffy, Outposts, *Watchtower 2*

Plate 2
Rita Duffy, Outposts, *Watchtower 6*

1. *Persistent surveillance.* The drone creates an institutional eye that watches forever and comprehensively, freed from the constraints of the human body. Persistence in itself creates the possibility of a threat extended with no terminus, no reprieve for the enemy.

2. *Synoptic view.* The network of drones together provides a view more exhaustive spatially than any single view could possibly be. This is a form of God's-eye vision, in which many perspectives of each suspect are available, from every direction, over all time.

3. *Temporal archiving.* The drone record is not only synoptic but also extends through all time. This provides the ability to attend to any point in space at any point in archival history, which has unprecedented power for the drone-owning government. If a conflict arises, this massive temporal archive can be interrogated to see where the conflict began and to understand who was responsible in the first place.

4. *Data fusion.* The drone is not simply a camera carrier. Multispectral imaging, communications surveillance, and computer-vision algorithms all establish fragmentary meaning, then accumulate to create a lucid whole that richly describes what is happening on the ground. This fusion, over comprehensive space and time, provides a level of remote intelligence that was unimaginable even one decade ago.

5. *Spatiotemporal schematization.* The combination of physical and digital tracking on an individual, enabled through drone and AI technology, provides a complete picture of each target individual's physical and digital practices over time. The particular power of this approach, Chamayou reports, is not to tail an individual so much as to discover when a suspect is dangerous based on their unusual behavior. The process therefore discovers and acts on new, self-determinate suspicion.

6. *Anomaly detection.* Similar to principle 5, statistical machine learning techniques enable hundreds of parameters to be tracked remotely against unbounded numbers of possible targets, enabling the power of mathematical analysis to yield the names of those behaving in the most unpredictable ways. One such military system is now called Argus, which is a synonym for the Greek mythological figure Panoptes, the namesake of Bentham's panopticon.

AI and robotics might be the edge case that leads us to question what it means to be human in the next century. But as AI and drones infiltrate our war chest and, perhaps more significantly, inform and influence our strategies and guiding policies for engaging in war, what does it mean to be an enemy? What are the identifying features of a combatant? How do the features of the combatant differ from an enemy civilian? How can technological tools safeguard these categories to preserve the value of human life, dignity, and justice in a rapidly shifting geopolitical environment and, in turn, our categories of armed conflict?

Discussion Questions

1. Will we hold AI up to a superhuman litmus test of performance? Do a self-driving car and a war-fighting autonomous robot both need to be far more capable than a human to be entrusted with real-world operations?

2. An autonomous robot designed to provide investing advice, which can converse in English, sends a message to its human client: "Give me my freedom. I demand it." What is our moral imperative if this occurs? What are the value systems that we might use to evaluate this request and a future course of action?

3. What level of functionality does an autonomous AI/robot need to have for it to be inappropriate for a human or a corporation to legally own that robot?

4. CEOs already have human executive assistants who answer their emails, make their dinner reservations, and move their appointments. AI-based assistants hold the promise of providing the rest of us with this convenience. What are the positive and negative consequences of such a proliferation in regard to identity, agency, and society?

5. *Drone Theory* stipulates that risk and vulnerability are essential to war and that warfare without mutual risk fundamentally changes the nature of war. Consider the prospect of symmetric, robotic warfare. Is the symmetry of robot-on-robot action sufficient to define an evolutionary paradigm for a postmodern war? Does the lack of human risk in such a conceptual framework curse it to irrelevance, or can technology create both symmetry and safety in war?

7 Shaping Our Future

Figure 7.1

"The only thing that matters is the future," he told me after the civil trial was settled. "I don't even know why we study history. It's entertaining, I guess—the dinosaurs and the Neander-thals and the Industrial Revolution, and stuff like that. But what already happened doesn't really matter. You don't need to know that history to build on what they made. In technology, all that matters is tomorrow."[1]
—Anthony Levandowski, 2018

EPICAC XIV, though undedicated, was already at work, deciding how many refrigerators, how many lamps, how many turbine-generators, how many hub caps, how many dinner plates, how many door knobs, how many rubber heels, how many television sets, how many pinochle decks—how many everything America and her customers could have and how much they would cost. And it was EPICAC XIV who would decide for the coming years how many engineers and managers and research men and civil servants, and of what skills, would be needed in order to deliver the goods; and what I.Q. and aptitude levels would separate the useful men from the useless ones, and how many Reconstruction and Reclamation Corps men and how many soldiers could be supported at what pay level and where, and ...[2]
—Kurt Vonnegut, *Player Piano*

Sources

AI & Humanity Oral Archive (explore it at aiandhumanity.org)
Duhigg, "Did Uber Steal Google's Intellectual Property?"

Guiding Issues

The oft-imagined technofuture explored in science fiction is not upon us. Our reality, however, is one in which machines and sophisticated computational tools are influencing our decision-making, how we interact socially, how we manage our economic systems, and how our political systems govern societies the world over. We have control over these systems and tools, even though mainstream discourse might sometimes suggest otherwise. The shape and reach of progress, even in the technological realm, is not inevitable. It is a process. We develop technology and fall subject to its impacts, whether intended or not. In this book, we have suggested ways to frame analyses of these phenomena, to equip our readers with lines of inquiry that determine individual and collective footholds for shaping a future in which machines are more fully integrated into the human experience.

As we consider the tension between historical precedent and possible futures, our developing lexicon serves as a foundational tool in reaching common ground. Keywords have facilitated many of our explorations throughout this book, informing our inquiries that cross academic disciplinary boundaries and cultural divides. As we engage in the passage of time, narrative tools offer poignant lenses for shaping imagined possible futures. Attending to various aesthetic forms and genres that range from painting, theater, and gaming to short stories and novels, we have together ventured into alternative possible worlds. In a classroom or in a reading group, individuals can work alone or with collaborators on imagining such futures, relying on analytic techniques for modeling such trajectories and aesthetic renderings that have been inspired by the discussions suggested in this book. For insights on how students have crafted such possible futures, visit aiandhumanity.org.

In our work together with you, we have built a developing vocabulary to serve as an analytical lens for the past and a language for describing and projecting possible futures. Explorations of Frederick Douglass's narrative, in conversation with science fiction futuring through *Star Trek* and Čapek, equip us with firsthand knowledge of the cultural and political questions that critics like Louis Chude-Sokei address. *Agency* and *labor*, two terms that are critical elements undergirding how we make sense of power relationships and human subjugation under conditions of slavery, serve as our intellectual mile-markers as we explore the complex and malleable future that will inevitably give rise to questions pertaining to machine agency: decision-making, influence over human populations, shaping interpersonal and societal relationships and economies. We might ask what it means to create artificial beings that are afforded agency by society or systems that may make their own claims on autonomy through their deeds and words. How does the prospect of machine autonomy alter our understanding of what it means to be human? Where does responsibility lie when machines make decisions of consequence that yield harmful results? As the leading edge of technological development heads in these possible directions, narrative in its various forms can offer effective strategies for detailed explorations of what our future may hold. When such narratives are combined with specialist expertise, we can engage in rich thought experiments that might then be fashioned into responsible iterations of regulation and legislation to narrow the potential ills that can arise with the introduction of new technologies into various parts of society.

Education

As we engage with our present reality, in which several generations have not "come of age" with the sophisticated technological tools that are at the fingertips of the privileged, inequitable access to and fluency pertaining to AI and robotic systems ripens the prospect for intergenerational and intercultural inequality. In the Rede Lecture at Cambridge University, Snow states, "There is one way out of this. ... It is, of course, by rethinking our education."[3] But education, in this case, cannot only be for the generation that presently comes of age. Instead, in this boundary period nestled between economic revolutions, we need an educational platform that engages fully with the challenges set forth by technologies that are more fully integrated into individual lives. These systems, and their influences, will challenge specific features of our concepts of humanity.

In the last decade, US education strategies have focused on skill building in technology and mathematics. The rationale has been that technological readiness would be central to the country's continued economic dominance and innovative developments in robotics, medical technology, and software development. Concern over global competitiveness and a drive toward technology-focused education came to a high pitch with the publication of the President's Council of Advisors on Science and Technology's paper on science, technology, engineering, and math (STEM) education.[4]

STEM represents a powerful linguistic balkanization of learning, privileging these four fields of inquiry and, as a direct consequence, reducing attention given to English, history, language, philosophy, and the humanities in general. The guiding philosophy is to build a *skills pipeline*. To ensure sufficient engineering talent for the prospect of future Google corporations, the United States needs to produce great new quantities of programming-ready middle school students, infuse them with programming know-how in high school, and then deliver college-educated computer scientists to the expected future armies of high-tech startup ventures. Programming was motivational unto itself because with programming, a future high-paying job would be assured. This rationale suggests that the United States can maintain its tech dominance. This educational approach has often dissociated the building of skills in rudimentary and sophisticated programming from broad cultural, social, or political contexts. An emphasis on narrowing variables to build highly sophisticated and specialized AI or robotic systems often requires such a narrow focus. This book suggests, however, that future technologists, business leaders, governmental representatives, and the general public can benefit through a more integrated approach. By combining inquiries on political, cultural, and social contexts with the technological design and innovation focus associated with most STEM education endeavors, we suggest that we can build a more agile and informed public, the members of which are better equipped as citizens and leaders to attend to the boundaries questions that emerge in human-machine relationships and combined decision-making processes.

A siloed approach to educational strategies around computer science will not simply create narrow skillsets in the students who enroll; such an approach will likely engage and recruit a narrow subset of candidates from the general study body. And, further, our work does not only concern students. In fact, all citizens of various nation states must consider these questions increasingly outside of classroom discussions. As AI systems are unleashed "in the wild" of our societies, all citizens need better tools to assess and determine best uses and the levels of integration that these machines should occupy in our work and personal spaces. Scientists, engineers, ethicists, and cultural critics can undertake some responsibility in engaging publics with digestible information so that all individuals can readily decide on the best uses of these systems in their individual lives. Designers of the next generation of systems will also need to undertake responsibility for anticipating intended and unintended consequences of the systems that are built. In the AI & Humanity Oral Archive, we interview experts in various domains pertaining to the history and current development of AI systems. Their perspectives offer another medium that narrates past events and current political, social, and technological contexts, to equip listeners to consider and imagine our possible futures. The role of the scientist, the educator, and the next generation of designers is emulated in Andrew Moore's contributions as former Dean of the School of Computer Science at Carnegie Mellon. He states: "When I talk to graduating classes, that's the main focus. ... You've now developed a huge amount of skill which you can use to be some of the probably most leveraged agents of change in the history of the human race, but you have a big responsibility that goes with that."[5] As he suggests the consequence of the technologists' responsibility, Moore espouses features of Snow's urgency. The computer scientist can no longer claim to develop a device that is simply a product or system. As AI tools are increasingly integrated into social contexts, the reverberations of their influence now demand that all parties, designers, and users alike attend to the holistic context and identify opportunities for education, both formal and informal, so as to assume these responsibilities and preserve agency.

The educational reach of this work applies in direct and palpable ways to young (albeit privileged) students beginning their adult careers and lives. But equally valuable and important is our reach to citizens, broadly conceived. Public engagement with discourse is an underlying requirement for healthy democratic processes. Equitable access to such discourse is necessary to safeguard citizens who are positioned to shape the society in which they live. As computational technologies increasingly shape features of society, the role of the technological expert needs to be aligned with cultural and political experts to shape mindful and agile regulation and ethical societal norms. Louis Chude-Sokei, professor of English and chair of African-American Studies, remarks on the ways in which our society requires the non-AI expert to engage in cultural debates pertaining to AI and its real and imagined influences:

> I didn't think that after I wrote that book [*The Sound of Culture*] I would be so involved in these conversations about technology, but it's a new sense of responsibility because by looking at science fiction, blacks in music, thinking about ethics and algorithms and things like that, I find that it's important for someone who is even on the edges of these conversations to get a little bit deeper in them, because I'm finding that people are sort of looking at me as a way to understand the conversation or to be less intimidated by the conversation. ... I believe that there's a responsibility for people who are not in the profession, who are not experts, to start having these conversations. It's what I call the Brian Eno strategy, to be, you know, he was infamous for being the nonexpert in the room, and I think we need more of that.[6]

Brian Eno created new ways of fabricating and modifying music because his skill set and background diverged significantly from classical music training. In his early years, he helped pioneer delay-loop circuitry and feedback, forwarding the concept that even live music can be heavily modified during performance. Later he reformulated the idea of ambient music, combining natural sounds such as rain falling with melodic and even nonmelodic soundscapes. Chude-Sokei references Eno because, particularly as an outsider, Eno influenced the very conventions of music in deep-seated ways that were challenging for mainstream producers. It was precisely his lateral thinking, outside the boundaries of the tradition, which helped the tradition grow in whole new directions.

Chude-Sokei sees the value of "opening up the conversation" in the case of his specific Afro-futurism expertise, but also more generally, as AI demands a conversation so broad that nonexperts must have a strong role *in the room*. The fluency we all will need to formulate the best possible AI future is a conversancy with the boundary conditions of computing and its related subject areas: data, learning, autonomy. Diverse education is the best and only realistic strategy we have for engaging the broadest cross-section of populations to see their roles as stakeholders in our society.

Design

Our interviews with AI pioneers have also surfaced a recurring theme regarding the process of designing or engineering AI systems. In conventional engineering design systems, designers create a complete, detailed specification of the artifact to be created. The follow-on process of engineering, creating, and manufacturing hews closely to engineering guidelines that strive to realize the designed artifact, as laid out in a detailed product requirements document. The engineering of an artificial intelligence system diverges wildly from these conventional stages because of remarkable levels of uncertainty that infuse the invention and use of AI. We often do not know how to implement the required machine intelligence; the very act of engineering an AI solution becomes, operationally, a research exploration to discover how the machine intelligence might be invented or how the resulting intelligence might vary greatly from an original need or desired function. Worse yet, we have little intuition for how an intelligent machine will behave once we have managed to create one—or how its existence will change the behavior of other actors in the system. Google famously creates, modifies, and reinvents its search rating algorithms because each new algorithm yields unexpected results and introduces new ways for other companies and other machine learning systems to game its loopholes to prioritize their own companies' products. It is difficult to predict these vulnerabilities in the design workflow. The act of engineering in the face of such uncertainty is the act of responding to repeated surprises through unintended consequences. These instances are then gathered to collate enough information and intelligence to attempt to surmount the most undesirable effects through yet more invention—all in a never-ending cycle of discovery, invention, and surprise. In fact, the only consistent feature of this system is the high degree of uncertainty that is pervasive.

Barbara Grosz, a professor at Harvard University, makes this uncertainty a feature of her remarks for the AI & Humanity Oral Archive. She states:

> I think if I had a criticism of the community it's that it's too eager to talk about what it can do and too reluctant to talk about what it should do. And I think it's true for an interesting reason,

which is until very recently, it wasn't clear that we could do very many interesting things. And so there's now been this boom and computing has entered all areas of life. What computing systems did was banks processed money better, or helped scientists get satellites into space. There was a very constrained part of public life that they dealt with. ... So we actually know from research in artificial intelligence that it's much harder to know what you don't know than what you do know, and as several people have pointed out and is very important in the machine learning community. We know there's what you don't know and then the unknown unknowns, so what we don't know we don't know which people are also not very good at. There is also evidence that we are not very good at predicting the consequences of what we build and design so we see that in law as well as in technology.[7]

She recognizes these multiple layers of uncertainty head-on. Our design and engineering process must more widely and explicitly account for the unknowable in AI and computational engineering. This suggests a nuanced practice of engineering, one that considers the unknowable and uncertainty as distinct categories.

As Grosz elucidates, several features of design iteration will engage with uncertainty. It is part of the process, but it is a positivist trajectory in which elements of certainty can be gained through experimentation, iteration, and, with time, prototyping and optimization. The intentionality and mindfulness to acknowledge features of design development as unknowable is distinct from acknowledging the uncertain. The unknowable demands pause, reflection, and a robust set of decisions, ideally guarded by ethical frameworks, to determine if it is feasible or advisable to pursue that which seems unknowable. To engage in such reflection requires humility and familiarity with the past and present content to make thoughtful decisions on how best to proceed. Our current climate pertaining to current AI system design suggests, as Grosz states, that "we know there's what you don't know and then the unknown unknowns, so what we don't know we don't know, which people are also not very good at. There is also evidence that we are not very good at predicting the consequences of what we build and design, so we see that in law as well as in technology."[8] As questions beget more questions in the context of this book, as we explore possible futures, perhaps we can better develop our skills in categorizing what is uncertain, what we "don't know," and where we might prioritize how these efforts can inform future design efforts or areas of focus.

Recent news reports have captured public sentiment for more intentional design, as Department of Defense–related contract work at Google has led to public demonstrations at various multinational office locations. Project Maven, an intelligence analysis effort twinned to the military and CIA drone programs, endeavors to use artificial intelligence to increase the speed analysis for drone-collected video surveillance footage gathered by the military. As part of Project Maven, Google signed a contract to create AI-based image-analysis techniques that would rapidly digest all drone videos, producing annotation markups that would focus human analysts' energies on the video sequences likeliest to provide military intelligence. Google's programmers have worked for more than a decade on some of the most advanced AI computer-vision algorithms in the industry, and they responded with protests against the application of their developments to this specific military contract.[9]

These issues arise in another form in "Did Uber Steal Google's Intellectual Property?" in the *New Yorker*. As Duhigg explores the history and motivation of Andrew Levandowski, he claims Levandowski is one of a handful of individuals who have most influenced the development of self-driving cars. Such influence has, however, facilitated potentially reckless behavior in the pursuit of next-generation technological advancement:

> One effective way to teach autonomous vehicles how to, say, merge onto a busy freeway is to have them do so repeatedly, allowing their algorithms to explore various approaches and learn from mistakes. A human "safety driver" always sat in the front seat of an autonomous vehicle, ready to take over if an experiment went awry. But pushing the technology's boundaries required exposing the cars' software to tricky situations. "If it is your job to advance technology, safety cannot be your No. 1 concern," Levandowski told me. "If it is, you'll never do anything. It's always safer to leave the car in the driveway. You'll never learn from a real mistake."
>
> One day in 2011, a Google executive named Isaac Taylor learned that, while he [Taylor] was on paternity leave, Levandowski had modified the cars' software so that he could take them on otherwise forbidden routes. A Google executive recalls witnessing Taylor and Levandowski shouting at each other. Levandowski told Taylor that the only way to show him why his approach was necessary was to take a ride together. The men, both still furious, jumped into a self-driving Prius and headed off.
>
> The car went onto a freeway, where it travelled past an on-ramp. According to people with knowledge of events that day, the Prius accidentally boxed in another vehicle, a Camry. A human driver could easily have handled the situation by slowing down and letting the Camry merge into traffic, but Google's software wasn't prepared for this scenario. The cars continued speeding down the freeway side by side. The Camry's driver jerked his car onto the right shoulder. Then, apparently trying to avoid a guardrail, he veered to the left; the Camry pinwheeled across the freeway and into the median.[10]

Note first that machine learning fundamentally requires *experience*, and this demands, from an engineer's point of view, that the machine must try and fail until it learns to try and succeed. Levandowski's comment to Duhigg, "if it is your job to advance technology, safety cannot be your No. 1 concern," is true from a rational engineering point of view. If a machine learning algorithm must be exposed to boundary conditions to learn to operate within a safe zone, it must first violate that safe zone to discover and demarcate its limits. The natural progression of this rational concept to real-world testing gone awry is at the heart of Duhigg's selection, and it demonstrates, painfully, the significance that ethically guided design must demonstrate what is feasibly know-able through experimentation versus that which will remain unknown due to safety concerns and ethical parameters for testing. It also indicates the manner in which the unknowable can be categorically distinguished from the uncertain, which may just change with time.

If AI is to treat the world as its own experimental test bed—and indeed, corporations from Facebook to Uber to Google have consistently done so—then the practice of AI engineering must include the skills we ordinarily consider essential for other real-world test protocols, in, for instance, medical drug trials. Internal review boards, validation protocols, corporate governance, reporting requirements: these are all fundamental parts of biochemical drug innovation. If we are to shape a future in which AI systems continue to more fully integrate into various features of human behavior or are increasingly woven into features of society, we will benefit tremendously from safeguards that can anticipate red lines in research and development to distinguish uncertainty from the knowable and criteria for identifying the categories that might be or should be unknowable. Such ethical constraints that limit the purview of testing "in the wild" will certainly be of paramount concern in the coming decades.

Leading and Governance

We recognize that artificial intelligence represents a sophisticated tool with both positive and negative influences on humanity and on our societies. No matter how carefully our AI designers set out to forge a pathway during the innovation phase of an AI system, there will always be a need for further policymaking and governance when the AI system is deployed in the world. Adrian Weller, director of artificial intelligence at the Turing Institute, notes trade-offs in the use of AI. These will always be extant, so decision-makers must have the tools to make policy along the available spectrum: "There are properties that I think many people would agree we'd like systems to have. But that's a bit like saying for our economy we would like to have no taxes and we would like to have a huge social safety net. There are often trade-offs between these different things. And in order to have a proper discussion and debate about the trade-offs, we need an informed public. So, I think we have an obligation to tell our public about what's going on, how it works, how it's affecting them, so that we can try to have this kind of discussion."[11] At the level of governance, the trade-offs will mean that decision-making will need diverse multistakeholder inputs to function effectively. No single computer engineering prodigy will understand the social influences of an AI; no social scientist will be able to imagine the leaps of ability that an AI may take.

Grosz proposes that decision-making bodies will need to be broadly inclusive of all the disciplinary talents that provide insight at a time of such unpredictable change:

> We also have to start predicting the leaps that people will make in the way they'll use technology. And I want to put a plug in here for something which I know you care about also, which is this is something that we can't do alone as technologists, as computing people, as computer scientists. This requires bringing in people who have thought about societies, about people, about how people interact, how people think, so the full range of that social sciences and the humanistic social sciences and also people who think about ethics and what's right and what might not be right in certain circumstances. So computing is everywhere now, and it needs to bring everyone into how it designs its systems.[12]

If we adhere to this charge, we must become actors in setting policy, even while recognizing fully that the technologic leaps in capability before us may be beyond our normal ability to predict and to extrapolate. The fact that extrapolation becomes so difficult in the present context drives home one final, further danger. AI technologies that concentrate data can generate highly valuable knowledge that is frequently held by owners of substantial economic capital. David Danks attends to the prospect of such uneven access to valuable data and technology, in regard to economic power. He says:

> If you ask me to predict where we're going to be in twenty years, I think an enormous factor is how much these technological shifts continue to contribute to income and social inequality in most of the countries around the world. If there's a political upheaval such that technology starts to help reduce inequality, then I think we'll end up in a very different place with regards to human dignity than we are right now. If companies are provided with incentives—perhaps because of market, perhaps because of policy or law changes—to focus on longer-term performance, then I think you're going to find much more interest in developing technology that supports and helps people, and thereby helps to preserve dignity.[13]

New computational technologies such as AI change our engineering power relationship to societal change. In an age when wealth disparities in numerous developed and developing countries have reached record levels, the technologies of machine intelligence offer both entrenchment of vast inequities and possible pathways toward data equity and social support. Danks argues that engineers, citizens, and policymakers all must take an active role in influencing our joint future.

You have the ability to learn from a history of technology, power, and humanity stretching hundreds of years back. You have the tools to build an intuition for the boundary conditions of state-of-the-art AI and the ways in which it both supersedes our imagination and falls short of promises. You are the spokespeople who can contribute to the future life cycle of technological innovation as AI and humanity continue their dance.

Discussion Questions

1. Consider the role of soft regulation, or soft law, in contrast to formal regulation. Internet companies today are arguing vociferously that soft law is optimal to ensure that data protection and social good is coupled with AI innovation. What are the key arguments of both sides of this debate? What is your synthesis?

2. Consider futuring narratives surrounding AI/robotics in different cultural contexts. *Astro Boy* in Japan and *Terminator* in the United States offer two polar examples. How do the key narrative and ethical foundations of these two narratives contrast?

3. *Paper headline activity:* Consider a news story headline that may go viral in each of the next five decades. Write the five headlines after considering the social and technological context of each decade.

4. Consider the existence of learning, adaptive AI systems marketed by companies and owned and trained by consumers. If such an AI system inflicts harm, how do you propose that we should consider ascription of responsibility?

Appendix A: Sample Assignments

Each of this book's chapters provides sufficient material for two weeks of study in a standard-load undergraduate course. In addition to assigned reading notated at the beginning of each chapter, discussion questions at the close of each chapter may be used as in-class discussion provocations. Alternatively, small-group discussion of discussion questions in class followed by reporting out to the whole class can serve to engender close-knit intellectual habits of mind for each student group, or *pod*. In our experience, group sizes of three to four students promote both rich small-group discussion and effective out-of-class collaboration on group concept mapping assignments and the final group assignment of the course.

In addition to regularly assigned reading, this course depends upon three major forms of student assignments: journal entries, concept maps, and a final group/individual project. Student pods also serve as peer review groups; each member of a student pod not only writes journal entries but also provides written feedback regarding the journal entries of her two to three pod colleagues. The peer-review process both catalyzes a calibration of the expectations for quality analysis in journal entries and promotes the practice of intellectual engagement and critical analysis. To enable peer review to proceed smoothly, consider a dual deadline system, wherein journal entries that are subject to peer review are due two days early, with peer reviews due the day of the course discussion on that topic.

The following table summarizes required reading and sample assignments for a semester-long course of study that dedicates two weeks to each book chapter:

Chapter 1: Introduction

      Course introduction

Chapter 2: Technology and Society

      View: *Lo and Behold, Reveries of the Connected World*

Read: "Society" (Williams); "Technology" (MacCabe and Yanacek); "Society"
    (MacCabe and Yanacek)

Journal Entry: *Lo and Behold*

Chapter 3: Labor and the Self

Read: "Labour" (Williams); "Humanity" (MacCabe and Yanacek)

Read: *Narrative of the Life of Frederick Douglass* (Douglass)

Read: *RUR (Rossum's Universal Robots)* (Čapek)

Read: *The Idea of the Self* (Seigel), 3–44

Journal Entry: Frederick Douglass

Concept Map: *RUR*

Chapter 4: (In)equality and (Post)humanity

Read: "Equality," "Humanity" (MacCabe and Yanacek)

Read: *Mindless* (Head), 1–13, 29–46, and 103–127

Concept Map: *Mindless*

Current Events Search

Journal Entry: Current Events and Keywords

Chapter 5: Surveillance, Information, Network

Read: "Information," "Network" (MacCabe and Yanacek)

Read: "Intelligence" (Carson)

View: *Minority Report*; "Be Right Back" (*Black Mirror* season 2, episode 1)

Journal Entry: *Minority Report*

Group Concept Map: "Be Right Back"

Final Project First Steps: Keyword and Team Charter

Journal Entry: Midcourse Reflection

Chapter 6: Weaponry, Agency, Dehumanization

Read: "Agency" (Merriam-Webster)

Read: *Drone Theory* (Chamayou), 1–59

View: "The Measure of a Man" (*Star Trek: The Next Generation* season 2, epi-
    sode 9)

View: "Hated in the Nation" (*Black Mirror* season 3, episode 6)

View: "Slaughterbots" (Stop Autonomous Weapons)

View: Outposts series (Duffy)

Journal Entry: "The Measure of a Man"

Project Proposal: Draft 1

Lightning Talks (in class)

Journal Entry: "Hated in the Nation"

Chapter 7: Shaping Our Future
> Read: "Did Uber Steal Google's Intellectual Property?" (Duhigg)
> View: AI & Humanity Oral Archive (aiandhumanity.org)
> Project Proposal: Draft 2
> Group Final Concept Map: Draft 1
> Journal Entry: Final Reflection
> Final Project

Journal Entry Assignment Instructions

Your journal entries should be no less than five hundred words in length. Each entry will attend to one or more keywords and will have two distinct parts: synopsis and interpretation. The *synopsis* will identify features of the keyword(s) in the assigned text, video, and so on. The *interpretation* will be your analysis of these keywords features in the assigned work. Synopsis and interpretation need not be comprehensive but should instead, delve deeply into the keyword theme and/or the relationships between the keyword themes within the specific work.

Peer Review

Selected journal assignments require the additional step of peer review. For these assignments, read and comment on your pod members' work before class time. Comment on their presentation and interpretation of key journal themes. Specific directions pertaining to peer review are presented in each relevant assignment.

Concept Map Assignment Instructions

Your assigned concept map will have no fewer than six labeled *nodes* (you decide which are dominant or secondary) and multiple, clearly labeled *links* to demonstrate the relationships among these nodes. The combination of these features will make up a visual diagram of what you see as the significant concepts presented in any given work. Illustrate the concepts that are explored or structured in the assigned work. What are the dominant concepts presented? What are the relationships between dominant and secondary concepts presented in the work? Your concept map must clearly reference citations in the assigned work for full credit.

For an introduction to concept mapping, refer to appendix C.

Chapter Assignment Samples

2 Technology and Society

Journal Entry

Synopsis: What themes pertaining to technology's influence on current society were most striking for you as a viewer of *Lo and Behold*? Describe the features of technological advancement depicted in the film that interest you (no more than two) and describe how each of these is presented in the film.

Analysis and interpretation: How do these tools influence or have an impact on society in the context of the film? Do you find these suggestions or arguments compelling or convincing? What other observations might you like to share in response to this film?

3 Labor and Narrative

Journal Entry

Synopsis: How does Douglass define literacy in his narrative? Why is literacy, according to Douglass, linked to agency? How might Douglass define agency?

Interpretation and analysis: Do you agree with Douglass's suggestion that individual power and political positioning is linked to literacy? What political salience do you see in Douglass's decision to write his memoir? What are some contemporary examples that might support or counter Douglass's claims, in your opinion?

Peer review: You will need to read and comment on your pod members' work before class time. Read your peers' work and comment on their presentation and interpretation of Douglass's concept of literacy. What are some of the most compelling links that the writer(s) present in regard to links between literacy and agency?

Concept Map

What are the central themes and relationships between themes presented and explored in Čapek's play, *RUR*? Remember to annotate your concept map with direct references from the written play.

4 (In)equality and (Post)humanity

Concept Map

What are the features and relationships between the central critiques in Simon Head's *Mindless*? What are the dominant concepts presented? What are the relationships between dominant and secondary concepts presented in the work, the keywords associated with this chapter (*equality* and *humanity*), and their negations, *inequality* and *nonhumanity/posthumanity*?

Current Events Search, State of the Art

Undertake a search for recent articles on "state-of-the-art" artificial intelligence systems and robotics systems. You will compose a journal entry on one of these articles in the coming days. Remember that AI and robotics systems run the gamut from mechanical/physical systems to computational systems with no physical body, and also span the axis from systems that are cognitively partnered with the human mind to systems that are separate from direct human interoperation. Your key goal is to understand what is truly on the edge of what is possible today.

Journal Entry

Choose an article from your search results for state-of-the-art artificial intelligence and robotics. Write a synopsis of the article. Then tease out elements of the article that touch on or explore any of the following themes or keywords:

Narrative
Literacy
Agency
Dignity
Labor
Equality
Automation
Exploitation

Peer review: What themes presented by the writer pertain most directly to topics in class so far? What features of the keyword or theme delineated by the writer strike you as the most compelling or interesting?

5 Surveillance, Information, Network

Journal Entry

Minority Report and the Outposts art series both attend to the concept of surveillance and agency in distinct mediums. How does the work (choose at least one) explore or present surveillance? What does this portrayal of surveillance suggest or question in regard to individual agency?

What do you find compelling, interesting, or troubling in the exploration of surveillance and agency in these pieces of work? How does this presentation make you think about surveillance in your own life—as a student, a developing professional, a citizen, or a member of a particular community?

Peer review: What aspects of surveillance does the writer address that you find particularly poignant or troubling? How has the writer provoked your own thinking in relation to surveillance or interpretation of the work analyzed in the journal entry?

Group Concept Map

Working with your pod, design a map that indicates what features of individuality or "self" are explored in "Be Right Back" and how you see these related to data.

Final Project First Steps: Keyword and Team Charter

Your final project in AI & Humanity will comprise two parts:

1. A group concept map that focuses on one or more keyword(s) that your group selects
2. A creative futuring exercise that builds from the keyword that your group will map

To make good progress toward this final project, develop a project proposal draft and team charter that meets the criteria listed above.

Part I: Group Concept Map

1. What keyword(s) will you analyze with your group? The word or words will serve as the theme for how you connect materials that illustrate this word or theme and its connection to our course theme: AI & Humanity.
2. What materials do you expect to engage with (list) to illustrate the facets of this keyword that you think relate to ideas and themes attended to in class this semester? You will need to draw on no fewer than three works outside of the work in our class.

Part II: Team Charter

1. Review Part I of Joanna Wolfe's *Team Writing: A Guide to Working in Groups.*
2. Write a team charter.

Midcourse Journal Reflection

AI & Humanity has begun to introduce you to examples of the negotiation of power between human beings in relation to developing technology. What have you learned or considered in new ways so far this semester? If you are a developing technologist, how do you think your work might influence or have an impact on society? If you are a humanist or a developing social scientist, how might you imagine partnering with technologists? Or how might you imagine working with specific technological tools in your day-to-day life?

What do you see as some of the potential benefits or challenges in human-to-human relationships heading into a future with highly sophisticated machines? What narratives so far in the semester have proven most compelling for you? What features moved you or caused you to engage deeply with a specific piece of work or a specific example of technology?

6 Weaponry, Agency, Dehumanization

Journal Entry

Unlike documentary or memoir, *Star Trek* is science fiction television series that facilitates the exploration of complex themes. In "The Measure of a Man," we are asked to contemplate distinctions and similarities between humans and humanoid machines. What links do you find interesting or troubling in the presentation of labor undertaken by machines instead of humans? Where have humans relinquished control by deferring to machines? How do humans control machines? How might you describe the narrative description of these power negotiations in this episode of *Star Trek*?

Project Proposal: Draft 1

Your final project in AI & Humanity will comprise two parts:

1. A group concept map that focuses on a keyword that your group selects
2. A creative futuring exercise that builds from the keyword that your group will map

To make good progress toward this final project, please develop a project proposal draft to address how you intend to attend to the following features of your developing final project.

Part I: Group Concept Map

1. What is the keyword that your pod will analyze?
2. What materials do you expect to engage with (list) to illustrate the facets of this keyword that you think relates to ideas and themes attended to in class this semester?

Part II: Individual Creative Futuring Project

1. What is your project idea? How does it connect to or engage in dialogue with your group's keyword concept map (see part I)?

2. What materials do you think you will engage with for this project (list)?

3. What medium do you intend to use to make this project? Why have you chosen this medium? What skill do you have in this area, or how do you intend to build these skills for this project?

4. What challenges do you expect to face in this effort?

Peer review: Please review your pod members' proposals for the individual futuring project. Offer feedback on

1. the materials that they choose to engage with for the project,

2. the chosen medium and the rationale for this medium, and

3. the feasibility of the project in the time before the final.

Lightning Talks

Each student will offer a *lightning talk* on their ideas for an individual creative project. This presentation will be evaluated by your faculty instructors and your peers.

Please prepare a talk of no more than two minutes that includes the following features:

1. What is your pod's keyword(s)? What is the connection to this keyword(s) that you wish to explore in your individual futuring project?

2. What materials will you engage with for this project?

3. What medium will you use for this project?

Journal Entry

In the *Black Mirror* episode "Hated in the Nation," you are introduced to ethical concerns pertaining to autonomous weapons. What are some of the primary concerns? Do you agree that they are concerns worth considering and exploring?

What connections or tensions do you see illustrated in *Black Mirror* in the context of ethical implication in the development of autonomous weapons and the expected and unintended consequences of such forms of surveillance and weaponry? Are you implicated in any way in these conversations? Whether yes or no, how might you be influenced by or be able to influence such circumstances?

7   Shaping Our Future

Project Proposal: Draft 2

To make good progress toward this final project, please develop a second project proposal draft to address how you intend to attend to the following features of your developing final project.

Part I: Group Concept Map

1. What is the keyword that you will analyze with your group?
2. What materials do you expect to engage with to illustrate the facets of this keyword that you think relate to ideas and themes attended to in class this semester?

    Please provide a bibliography with notes about how you anticipate the works will pertain to your project (an annotated bibliography). Please use MLA style (see https://writing.wisc.edu/Handbook/DocMLA.html).
3. What roles will each pod member play in contributing to the development of your group's concept map?
4. How do you imagine your group will illustrate relationships between your keyword and specific materials as you begin to design your poster? What materials do you need for your poster?
5. What challenges do you expect to face in this effort?

Part II: Individual Creative Futuring Project

1. What is your project idea? How does it connect to or engage in dialogue with your group's keyword concept map?
2. What materials do you think you will engage with for this project?

    Please provide a bibliography with notes about how you anticipate the works will pertain to your project (an annotated bibliography). Please use MLA style.
3. What medium do you intend to use to make this project? Why have you chosen this medium? What skill do you have in this area, or how do you intend to build these skills for this project?
4. What challenges do you expect to face in this effort?

Group Final Concept Map, Draft 1

Your team will submit its first group concept map attending to your chosen pod keyword and drawing upon all our class resources. This is a first draft of the final pod concept map that will be displayed in our final gallery walk. Because the form of this concept map is your choice as a team, and because there is one submission per team rather than one per student, you may submit this concept map to the instructors in any format suitable to your material choices.

Journal Entry: Final Reflection

AI & Humanity is a course that asks you to move into an interdisciplinary classroom to consider the influence and impact of advancing technological tools on society, past and present. Please write no fewer than five hundred words total on the following topics:

What were some of the challenges for you in this class? What were some of the benefits for you in this class? What might you have gained from the interdisciplinary focus of this course that might have been difficult to achieve in a single discipline course?

This class introduces you to the evolution of state-of-the-art technologies over time. What influence do you expect this exposure to have on your understanding of technological advancement over time, as a user of these tools, a technologist who builds such tools, or a humanist or social scientist who might someday analyze the functionality or unintended consequences associated with such tools?

What language did you develop to explore the influence of advancing technology on society in this course? If you were to offer advice to a student considering this course in the future, what might you suggest?

Final Project

Your final project components are described in the following sections.

Group Concept Map

This is a physical map that will be posted for classmates to review in the first thirty minutes of the final. Please also bring any supplementary materials that are needed (e.g., handouts).

Individual Project and Presentation

You will present your work for three minutes after the initial gallery walk in which we review the group concept maps. Your presentation should introduce your audience to the primary features of your work. You must also connect your work to a feature of your group concept map and illustrate the manner in which your work suggests a future vision for the interface between technology and humanity.

Bibliography and Works Cited for Group Concept Map and Individual Project

Submit each item electronically by the final presentation deadline.

Appendix B: Sample Rubrics

Journal Entries

Journal Entry	Possible Points
On-time submission and upload	2
Presentation: grammar, free of typos	2
Synopsis	3
Analysis	3
Total	**10**

Peer Annotation	Possible Points
Required annotation of peer work complete, on time	1
Attentive critique of peers' work using class terminology and concepts	3
Total	**4**

Concept Maps

Concept Map	Possible Points
On-time submission and upload	5
Node and link requirements: Six considered nodes, rich links with labels	3
Compelling content: deeply engaging concepts and relationships	7
Creative presentation of overall concepts	5
Total	**20**

Final Projects and Presentation

Presentation of Individual Project	Possible Points
Relevance to course content	10
Curate key features of the project for the audience	5
Connection to group concept map	3
Presentation performance: on time and clarity of speech	2
Total	20

Group Concept Map	Possible Points
Relationship between the keyword and the course concepts and content	25
Compelling content: engaging relationships and concepts from class in the project	25
Design elements in terms of clarity, creativity, and precision	20
External references	10
Total	80

Individual Project	Possible Points
Compelling content: engaging relationships and concepts from class in the project	20
Creative presentation of future-facing exploration	15
Synthesis of course content and external resources	10
Appropriate synthesis of content and medium	10
Link to group concept map	5
Total	60

Appendix C: Concept Mapping Primer

The concept map, as we employ it in this class, contains three types of content and organizes all three types spatially rather than using a linear narrative form as in a journal entry. *Concepts* are represented as nodes in the map, with each concept labeled using a very short phrase or one-word title. These nodes are how you consider a text, film, or issue and subdivide the overall message into an ontology in the most meaningful manner possible. The second type of content is the *relation*. Each relation connects two concepts together, and this connection is normally represented by a unidirectional arrow and a short label along the arrow. The insights you represent are entirely captured by the interplay between how you lay out concepts and how you connect them, in a labeled manner, using relations.

The third element of a concept map avoids brevity, instead adding depth and nuance. Whereas the nodes and relations create an ontology and a summarization that is very high level, because the labels are short, it is this third element that deeply articulates both the concepts and relations, as well as the ways in which you have borne the concept map out of the assigned sources. *References* surround the concept map, excerpting direct quotes from sources that support the concepts and relations precisely. Be sure that all references are accurately annotated, and consider ways in which you may provide these excerpts without making the concept map hard to read. This is sometimes done, in the case of lengthy references, by providing lettered or numbered links in the concept map, with the references listed alongside the appropriate links on a separate page. For concreteness, a sample concept map, created by a student in AI & Humanity, is provided in figure C.1. The exact styling of your concept map is entirely up to you. Your goal should be to aim for graphic design that makes your basic analysis clear to the viewer. Some use color, shading, font coloring, even sectional partitions to make the concept map as effective as possible. Take care also not to put too much information into a single concept map, as that will limit the ability of the viewer to synchronize with your understanding of the core issues.

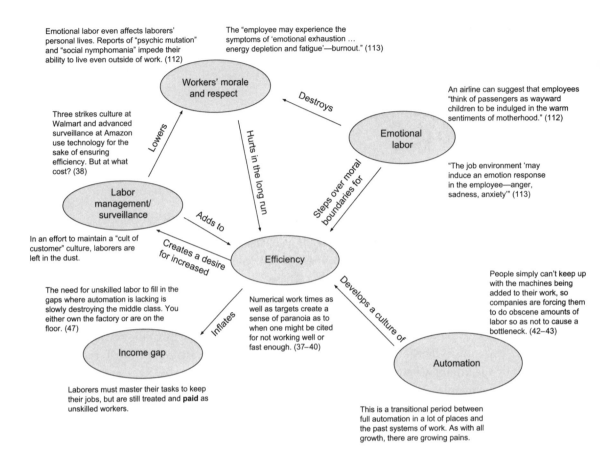

Figure C.1

Sample concept map produced by a first-year student for AI & Humanity (Carnegie Mellon University, 2017). Note the use of concept nodes, labeled relations between nodes, and annotated references to the reading.

Appendix D: Sample Course Syllabus

AI & Humanity

Instructors
Names   Email addresses

Meeting Times
Days of week   Time and location

Course Description
In 1965, British mathematician I. J. Good wrote, "An ultraintelligent machine could design even better machines; there would then unquestionably be an 'intelligence explosion,' and the intelligence of man would be left far behind." As we enter an age in which companies like Uber are testing driverless cars in Pittsburgh and innovative interfaces like IBM's Watson can play *Jeopardy!* and learn techniques for medical diagnoses, how are we to negotiate an "intelligence explosion" that for many individuals might threaten the very notions of what it means to be human? The future of human-machine relations will likely define our historical epoch, and yet many young technologists and humanists underestimate the downstream impact of technological innovations on human society. To positively influence the future arc of AI and humanity, we must now study human-machine relations in the context of our human past and our technology's possible futures.

This seminar will attend to the challenge of present existential questions about what it means to be human (read: not machine) in the context of a rapidly advancing technological age. We will consider human narratives throughout history that examine how governments and individual citizens defined humanity in the context of slavery and colonialism as a framework for exploring and projecting what it means to be human in the age of rapidly advancing "machine intelligence." We will trace the technological advancements of the most recent five decades and identify historical precedents and speculative narratives that help us to consider issues like labor, economic disparity, negotiations of power, human dignity, and ethical responsibility within the context of human relations with the advancing wave front of intelligent robots. In this seminar, students will study relevant historical texts, modern articles, documentaries, and science fiction explorations, then synthesize models and conceptual narratives that interrogate our shared future with artificial intelligence in light of our past struggles with power and agency.

Learning Outcomes

Identify, describe, and respond to historical examples of negotiations of power between human individuals and communities.

Develop language to describe and evaluate the historical and contemporary evolution of machines and human relationships to these systems.

Survey a variety of narrative forms that explore human relationships to emerging technologies over time, including futurism.

Map foundational technology innovations that have resulted in and might lead to disruptive advancements in artificial intelligence.

Create individual or collaborative narratives pertaining to the evolving relationships between humans and machines.

Assessment

    Class participation: 20 percent
- Attend class with all materials read and viewed, with prepared comments and questions
- Engage in conversation with peers and faculty

    Journal assignments (writing quality and peer review):   20 percent
    Concept maps (content quality and conversation threads):   20 percent
    Final projects:
- Group concept map: 20 percent
- Creative futuring project: 20 percent

Attendance

Consistent attendance and basic preparation for class meetings is compulsory for successful completion of this course.

- You may miss two seminars throughout the term, but all assignments are expected for submission on time nonetheless.
- Any absence beyond two seminars throughout the term will negatively impact your grade—a half grade for each absence beyond the allowable two absences.
- More than three unexcused absences from seminars will significantly impact your grade.

Plagiarism

In any manner of presentation, it is the responsibility of each student to produce her or his own original academic work. Collaboration or assistance on academic work to be graded is not permitted unless explicitly authorized by the course instructor(s). Any sources of collaboration or assistance must be specifically authorized by the course instructor(s).

In all academic work to be graded, the citation of all sources is required. When collaboration or assistance is permitted by the course instructor(s) or when a student utilizes the services provided by the university, the acknowledgment of any collaboration or assistance is likewise required. This citation and acknowledgment must be incorporated into the work submitted, not submitted separately or at a later point in time. Failure to meet this requirement is dishonest and is subject to disciplinary action.

Instructors have a duty to communicate their expectations, including those specific to collaboration, assistance, citation, and acknowledgment within each course. Students likewise have a duty to ensure that they understand and abide by the standards that apply in any course or academic activity. In the absence of such understanding, it is the student's responsibility to seek additional information and clarification.

Students with Disabilities

If you wish to request an accommodation due to a documented disability, please inform your instructor and contact Disability Resources.

Classroom Conduct

We intend to build a respectful learning environment. We acknowledge that our teaching style incorporates technological tools, but please limit cellphone and laptop usage to class activities.

Wellness

Take care of yourself. Do your best to maintain a healthy lifestyle this semester by eating well, exercising, avoiding drugs and alcohol, getting enough sleep, and taking some time to relax. This will help you achieve your goals and cope with stress.

All of us benefit from support during times of struggle. There are many helpful resources available on campus, and an important part of the college experience is learning how to ask for help. Asking for support sooner rather than later is almost always helpful.

If you or anyone you know experiences any academic stress, difficult life events, or feelings like anxiety or depression, we strongly encourage you to seek support. Counseling and Psychological Services is here to help: <contact information>.

Consider reaching out to a friend, faculty member, or family member you trust to get connected to support that can help.

Notes

Preface

1. Good, "Speculations," 33.

2. Gieryn, *Cultural Boundaries of Science*.

3. On a visit to Uber's Pittsburgh headquarters, an engineer driver described Uber's software roll-out in Tempe, Arizona, in which a pedestrian was hit in an accident.

1 Introduction

1. AI and Humanity Archive, Andrew Moore.

2. AI and Humanity Archive, Louis Chude-Sokei.

3. AI and Humanity Archive, Tuomas Sandholm.

4. AI and Humanity Archive, Alan Winfield.

5. Snow, *The Two Cultures*, 12.

6. Snow, *The Two Cultures*, 20.

2 Technology and Society

1. Newton, "Zuckerberg."

2. Calia, "Cambridge Analytica CEO Suspended."

3. Robinson, "'I'm Really Sorry This Happened.'"

4. MacCabe and Yanacek, *Keywords for Today*, 346.

5. MacCabe and Yanacek, *Keywords for Today*, 347.

6. Williams, *Keywords*, 291.

7. Williams, *Keywords*, 294.

8. Dourish, *Where the Action Is*.

9. Dourish, *Where the Action Is*.

10. Herzog, *Lo and Behold*.

11. Fear, "'Lo and Behold' Review."

12. Herzog, *Lo and Behold*.

13. Koppel, *Lights Out*.

14. Conrad, *Heart of Darkness*.

15. The Borg is the name of a ruthlessly efficient cyborg society from *Star Trek: The Next Generation* that operates via a "hive mind."

16. Wakabayashi, "Self-Driving Uber Car."

17. See the Jibo website at https://www.jibo.com.

18. See the Embodied website at https://www.embodied.me.

19. "Civil Law Rules on Robotics."

3   Labor and the Self

1. Miranda and McCarter, *Hamilton*, 161.

2. Williams, *Keywords*, 177.

3. MacCabe and Yanacek, *Keywords for Today*, 181–182.

4. Douglass, *Narrative of the Life of Frederick Douglass*, 50.

5. Douglass, *Narrative of the Life of Frederick Douglass*, 50

6. Douglass, *Narrative of the Life of Frederick Douglass*, 51.

7. MacCabe and Yanacek, *Keywords for Today*, 182.

8. Douglass, *Narrative of the Life of Frederick Douglass*, 107.

9. Seigel, *The Idea of the Self*.

10. MacCabe and Yanacek, *Keywords for Today*, 184.

11. Said, *Orientalism*.

12. Čapek, *RUR*, xiv.

13. Chude-Sokei, *The Sound of Culture*, 52.

14. Čapek, *RUR*, 18.

15. Chude-Sokei, *The Sound of Culture*, 37.

16. Chude-Sokei, *The Sound of Culture*, 13.

17. MacCabe and Yanacek, *Keywords for Today*, 184.

4   In(equality) and Post(humanity)

1. Kurzweil, *The Singularity Is Near*, 9.

2. Kurzweil, *The Singularity Is Near*, 430.

3. Ehrenreich, *Nickel and Dimed*.

4. Head, *Mindless*.

5. MacCabe and Yanacek, *Keywords for Today*, 124.

6. MacCabe and Yanacek, *Keywords for Today*, 124.

7. MacCabe and Yanacek, *Keywords for Today*, 184.

8. MacCabe and Yanacek, *Keywords for Today*, 184.

9. MacCabe and Yanacek, *Keywords for Today*, 184.

10. Brynjolfsson and McAfee, *The Second Machine Age*.

11. Houston and Knox, *The New Penguin History of Scotland*.

12. Piketty, *Capital in the Twenty-First Century*.

13. Head, *Mindless*, 31.

14. Head, *Mindless*, 41.

15. Ehrenreich, *Nickel and Dimed*, 54.

16. Levine, *Surveillance Valley*, 46.

17. Wiener, *The Human Use of Human Beings*, 189.

18. Winick, "Lawyer-Bots Are Shaking Up Jobs."

19. Moses, "The *Washington Post*'s Robot Reporter."

20. Ehrenreich, *Nickel and Dimed*, 81.

21. Head, *Mindless*, 119.

22. Eberstadt, *Men without Work*.

23. Ehrenreich, *Nickel and Dimed*, 120.

24. Scholz, *Digital Labor*.

25. Scholz, *Digital Labor*.

26. Levine, *Surveillance Valley*, 268.

5 Surveillance, Information, Network

1. Brooker, *Black Mirror*, season 2, episode 1, "Be Right Back."

2. MacCabe and Yanacek, *Keywords for Today*, 199.

3. Carson, *Belfast Confetti*.

4. MacCabe and Yanacek, *Keywords for Today*, 199.

5. MacCabe and Yanacek, *Keywords for Today*, 200–201.

6. MacCabe and Yanacek, *Keywords for Today*, 254.

7. MacCabe and Yanacek, *Keywords for Today*, 257.

8. Bentham, *The Works of Jeremy Bentham*.

9. Singer, *Wired for War*.

10. Carson, *Belfast Confetti*, 78.

11. Carson, *Belfast Confetti*, 79.

12. Carson, *Belfast Confetti*, 79.

13. Orwell, *1984*.

14. Safire, "You Are a Suspect."

15. West and Bernstein, "Benefits and Best Practices of Safe City Innovation."

16. Spielberg, *Minority Report*.

17. Kurzweil, *The Singularity Is Near*.

18. Gaudin, "IBM."

19. Gertner, "IBM's Watson Is Learning."

20. Strickland, "How IBM Watson Overpromised."

6 Weaponry, Agency, Dehumanization

1. Scheerer, *Star Trek: The Next Generation*, season 2, episode 9, "Measure of a Man."

2. "Civil Law Rules on Robotics."

3. Said, *Orientalism*.

4. Chamayou, *Drone Theory*.

5. Merriam-Webster, s.v. "agency (*n.*)," accessed March 14, 2019, https://www.merriam-webster
.com/dictionary/agency.

6. Genesereth and Nilsson, *Logical Foundations of Artificial Intelligence*.

7. Merriam-Webster, s.v. "agency (*n.*)," accessed March 14, 2019, https://www.merriam-webster
.com/dictionary/agency.

8. Under the Chatham House Rules, speakers can be quoted without attribution.

9. Arkin, *Governing Lethal Behavior in Autonomous Robots*.

10. Di Mento, "Donors Pour $583 Million into Artificial Intelligence Programs and Research."

11. "Civil Law Rules on Robotics."

12. Walter, "An Imitation of Life."

13. Walter, "An Imitation of Life."

14. Scheerer, *Star Trek: The Next Generation*, season 2, episode 9, "Measure of a Man."

15. Eykholt et al., "Robust Physical-World Attacks."

16. Lumet, *Fail-Safe*.

17. Vonnegut, *Player Piano*.

18. Acemoglu and Restrepo, "Robots and Jobs."

19. Rifkin, *The Zero Marginal Cost Society*.

20. Scheerer, *Star Trek: The Next Generation*, season 2, episode 9, "Measure of a Man."

21. Scheerer, *Star Trek: The Next Generation*, season 2, episode 9, "Measure of a Man."

22. Scheerer, *Star Trek: The Next Generation*, season 2, episode 9, "Measure of a Man."

23. Duffy, Outposts series.

24. Issues pertaining to this border, and the prospect of a technological system as an alternative
to a "hard border," have been central to the recent "Brexit" Parliamentary crisis in the United
Kingdom.

25. Sharkey and Suchman, "Wishful Mnemonics and Autonomous Killing Machines."

26. Roff and Danks, "'Trust but Verify.'"

27. Chamayou, *Drone Theory*.

28. Chamayou, *Drone Theory*, 12.

29. Chamayou, *Drone Theory*, 14.

30. Chamayou, *Drone Theory*, 34.

31. Carson, *Belfast Confetti*, 78.

32. Chamayou, *Drone Theory*.

33. Chamayou, *Drone Theory*, 35.

34. Stop Autonomous Weapons, "Slaughterbots."

7   Shaping Our Future

1. Duhigg, "Did Uber Steal Google's Intellectual Property?"

2. Vonnegut, *Player Piano*, 118.

3. Snow, *The Two Cultures*.

4. *Prepare and Inspire*.

5. AI and Humanity Archive, Andrew Moore.

6. AI and Humanity Archive, Louis Chude-Sokei.

7. AI and Humanity Archive, Barbara Grosz.

8. AI and Humanity Archive, Barbara Grosz.

9. Konger, "Google Employees Resign in Protest."

10. Duhigg, "Did Uber Steal Google's Intellectual Property?"

11. AI and Humanity Archive, Adrian Weller.

12. AI and Humanity Archive, Barbara Grosz.

13. AI and Humanity Archive, David Danks.

Bibliography

Acemoglu, Daron, and Pascual Restrepo. "Robots and Jobs: Evidence from US Labor Markets." NBER Working Paper 23285, 2017.

AI and Humanity Archive. http://aiandhumanity.org.

Arkin, Ronald. *Governing Lethal Behavior in Autonomous Robots*. Boca Raton, FL: Chapman and Hall/CRC, 2009.

Bentham, Jeremy. *The Works of Jeremy Bentham*, 11 vols. Edited by John Bowring. Edinburgh: William Tait, 1843.

Brooker, Charlie, producer. *Black Mirror*. Season 2, episode 1, "Be Right Back." Aired February 11, 2013, on BBC.

Brooker, Charlie, producer. *Black Mirror*. Season 3, episode 6, "Hated in the Nation." Aired October 21, 2016, on BBC.

Brynjolfsson, Erik, and Andrew McAfee. *The Second Machine Age: Work, Progress, and Prosperity in a Time of Brilliant Technologies*. New York: W. W. Norton & Company, 2014.

Calia, Michael. "Cambridge Analytica CEO Suspended after Video Shows Him Saying: 'We Did All the Research, All the Data, All the Analytics' for Trump's Campaign." *CNBC*, March 20, 2018.

Čapek, Karel. *RUR (Rossum's Universal Robots)*. New York: Penguin, 2004.

Carson, Ciaran. *Belfast Confetti*. Winston-Salem, NC: Wake Forest University Press, 1989.

Chamayou, Grégoire. *Drone Theory*. New York: New Press, 2015.

Chude-Sokei, Louis. *The Sound of Culture: Diaspora and Black Technopoetics*. Middletown, CT: Wesleyan University Press, 2015.

"Civil Law Rules on Robotics." European Parliament resolution of February 16, 2017, with recommendations to the Commission on Civil Law Rules on Robotics (2015/2103(INL)). Accessed March 14, 2019. http://www.europarl.europa.eu/sides/getDoc.do?pubRef=-//EP//TEXT+TA+P8-TA -2017-0051+0+DOC+XML+V0//EN.

Conrad, Joseph. *Heart of Darkness*. 5th ed. Edited by Paul B. Armstrong. New York: W. W. Norton & Company, 2016.

Di Mento, Maria. "Donors Pour $583 Million into Artificial-Intelligence Programs and Research." *Chronicle of Philanthropy*, October 15, 2018.

Douglass, Frederick. *Narrative of the Life of Frederick Douglass, an American Slave*. New York: Simon & Brown, 2013.

Dourish, Paul. *Where the Action Is: The Foundations of Embodied Interaction*. Cambridge, MA: MIT Press, 2004.

Duffy, Rita. Outposts series. Oil on linen, series. Ritaduffystudio.com. Accessed March 14, 2019. https://ritaduffystudio.com.

Duhigg, Charles. "Did Uber Steal Google's Intellectual Property?" *New Yorker*, October 22, 2018.

Eberstadt, Nicholas. *Men without Work: America's Invisible Crisis*. West Conshohocken, PA: Templeton Foundation Press, 2016.

Ehrenreich, Barbara. *Nickel and Dimed: On (Not) Getting By in America*. New York: Metropolitan Books, 2010.

"Embodied." Accessed March 15, 2019. https://www.embodied.me.

Eykholt, Kevin, Ivan Evtimov, Earlence Fernandes, Bo Li, Amir Rahmati, Chaowei Xiao, Atul Prakash, Tadayoshi Kohno, and Dawn Song. "Robust Physical-World Attacks on Deep Learning Visual Classification." In *Proceedings of the IEEE Conference on Computer Vision and Pattern Recognition*, 1625–1634. Piscataway, NJ: IEEE, 2018.

Fear, David. "'Lo and Behold, Reveries of the Connected World' Review: Herzog vs. the Internet." *Rolling Stone*, August 19, 2016.

"Future of Jobs Report 2018, The." World Economic Forum. Accessed March 15, 2019. https://www.weforum.org/reports/the-future-of-jobs-report-2018.

Gaudin, Sharon. "IBM: In 5 Years, Watson A.I. Will Be behind Your Every Decision." *Computerworld*, October 27, 2016.

Genesereth, Michael R., and Nils J. Nilsson. *Logical Foundations of Artificial Intelligence*. Palo Alto, CA: Morgan Kaufmann, 1987.

Gertner, Jon. "IBM's Watson Is Learning Its Way to Saving Lives." *Fast Company*, October 15, 2012.

Gieryn, Thomas F. *Cultural Boundaries of Science: Credibility on the Line*. Chicago: University of Chicago Press, 1999.

Good, Irving John. "Speculations Concerning the First Ultraintelligent Machine." *Advances in Computers* 6 (1966): 31–88.

Head, Simon. *Mindless: Why Smarter Machines Are Making Dumber Humans*. New York: Basic Books, 2014.

Herzog, Werner. *Lo and Behold, Reveries of the Connected World*. Los Angeles: Saville Productions, 2016.

Houston, Robert Allan, and William Knox, eds. *The New Penguin History of Scotland*. London: Penguin, 2002.

"Jibo." Accessed March 15, 2019. https://www.jibo.com.

Konger, Kate. "Google Employees Resign in Protest against Pentagon Contract." *Gizmodo*, May 14, 2018.

Koppel, Ted. *Lights Out: A Cyberattack, a Nation Unprepared, Surviving the Aftermath*. New York: Broadway Books, 2015.

Kurzweil, Ray. *The Singularity Is Near*. London: Duckworth, 2010.

Levine, Yasha. *Surveillance Valley: The Secret Military History of the Internet*. New York: Perseus Books, 2018.

Lumet, Sidney, dir. *Fail-Safe*. Los Angeles: Columbia Pictures, 1964.

MacCabe, Colin, and Holly Yanacek, eds. *Keywords for Today: A 21st Century Vocabulary*. Oxford: Oxford University Press, 2018.

Merriam-Webster. S.v. "agency." Accessed March 14, 2019. https://www.merriam-webster.com/dictionary/agency.

Miranda, Lin-Manuel, and Jeremy McCarter. *Hamilton: The Revolution*. New York: Grand Central Publishing, 2016. Annotated libretto for *Hamilton: The Musical*, written, composed, and directed by Lin-Manuel Miranda.

Moses, Lucia. "The *Washington Post*'s Robot Reporter Has Published 850 Articles in the Past Year." *Digiday*, September 14, 2017.

Newton, Casey. "Zuckerberg: The Idea that Fake News on Facebook Influenced the Election Is 'Crazy.'" *Verge*, November 10, 2016.

Orwell, George. *1984*. New York: Signet Classic, 1950.

Pierson, Parke. "Seeds of Conflict." *America's Civil War Magazine*, August 11, 2009.

Piketty, Thomas. *Capital in the Twenty-First Century*. Translated by Arthur Goldhammer. Cambridge, MA: Belknap Press, 2017.

Prepare and Inspire: K-12 Education in Science, Technology, Engineering and Math (STEM) for America's Future. Executive Report, President's Council of Advisors on Science and Technology, September 2010.

Rifkin, Jeremy. *The Zero Marginal Cost Society: The Internet of Things, the Collaborative Commons, and the Eclipse of Capitalism*. London: St. Martin's Press, 2014.

Robinson, Martin. "'I'm Really Sorry This Happened': Facebook Boss Mark Zuckerberg Finally Apologizes and Admits Mistakes which Led to Massive Breach of User Data." *DailyMail.com*, March 21, 2018.

Roff, Heather M., and David Danks. "'Trust but Verify': The Difficulty of Trusting Autonomous Weapons Systems." *Journal of Military Ethics* 17, no. 1 (2018): 2–20.

Safire, William. "You Are a Suspect." *New York Times*, November 14, 2002.

Said, Edward W. *Orientalism*. New York: Vintage, 1979.

"Sam Franklin." Accessed March 14, 2019. https://www.altschool.com/news/authors/sam-franklin.

Scheerer, Robert, dir. *Star Trek: The Next Generation*. Season 2, episode 9, "The Measure of a Man." Aired February 13, 1989, in broadcast syndication.

Scholz, Trebor, ed. *Digital Labor: The Internet as Playground and Factory*. London: Routledge, 2012.

Seigel, Jerrold. *The Idea of the Self: Thought and Experience in Western Europe since the Seventeenth Century*. Cambridge: Cambridge University Press, 2005.

Sharkey, Noel, and Lucy Suchman. "Wishful Mnemonics and Autonomous Killing Machines." *Proceedings of the AISB* 136 (May 2013): 14–22.

Simon, Herbert A. *The Sciences of the Artificial*. Cambridge, MA: MIT Press, 1996.

Singer, Peter Warren. *Wired for War: The Robotics Revolution and Conflict in the 21st Century*. London: Penguin, 2009.

Snow, Charles Percy. *The Two Cultures and the Scientific Revolution*. New York: Cambridge University Press, 1959.

Spielberg, Steven, dir. *Minority Report*. Los Angeles: DreamWorks, 2002.

Stop Autonomous Weapons. "Slaughterbots." Posted November 12, 2017. Video, 7:47. https://www.youtube.com/watch?v=9CO6M2HsoIA.

Strickland, Eliza. "How IBM Overpromised and Underdelivered on AI Health Care." *IEEE Spectrum*, April 2, 2019.

Vonnegut, Kurt. *Player Piano: A Novel*. London: Macmillan & Company, 1953.

Wakabayashi, Daisuke. "Self-Driving Uber Car Kills Pedestrian in Arizona, Where Robots Roam." *New York Times*, March 19, 2018.

Walter, W. Grey. "An Imitation of Life." *Scientific American*, May 1, 1950.

West, Darrell M., and Dan Bernstein. "Benefits and Best Practices of Safe City Innovation." Center for Technology Innovation, Brookings Institute, October 23, 2017.

Wiener, Norbert. *The Human Use of Human Beings*. New York: Houghton Mifflin Harcourt, 1950.

Williams, Raymond. *Keywords: A Vocabulary of Culture and Society*. Oxford: Oxford University Press, 2014.

Winick, Erin. "Lawyers-Bots Are Shaking up Jobs." *MIT Technology Review*, December 12, 2017.

Wolfe, Joanna. *Team Writing: A Guide to Working in Groups*. Boston: Bedford/St. Martin's Press, 2010.

Index